SAFEGUARDING THE UNIQUENESS OF THE COLORADO PLATEAU
An Ecoregional Assessment of Biocultural Diversity

CONTRIBUTORS

Tony Joe
Center for Sustainable Environments, Northern Arizona University

Luisa Maffi
President, Terralingua: Partnerships for Linguistic and Biological Diversity

Gary Paul Nabhan
Director, Center for Sustainable Environments at Northern Arizona University

Patrick Pynes
Terralingua Liaison and Center for Sustainable Environments, Northern Arizona University

David Seibert
Center for Sustainable Environments, Northern Arizona University

Thomas D. Sisk
Associate Professor, Envirnomental Sciences, Northern Arizona University

Lawrence E. Stevens
Grand Canyon Wildlands Council

Stephen Trimble
Author, Advisory Council member, Center for Sustainable Environments

SAFEGUARDING THE UNIQUENESS OF THE COLORADO PLATEAU
An Ecoregional Assessment of Biocultural Diversity

The Center for Sustainable Environments
Northern Arizona University, Flagstaff

Terralingua: Partnerships for Linguistic and Biological Diversity
Washington, D.C.

Grand Canyon Wildlands Council
Flagstaff, Arizona

With support from

The Ford Foundation
Agnese Haury
The Town Creek Foundation

ISBN 0-9718786-1-7

EXECUTIVE SUMMARY

This ecoregional assessment focuses on the unique biological and cultural resources of the Colorado Plateau in the southwestern United States, the threats to the region's astounding levels of biocultural diversity, and the possible means to safeguard or restore them. The Colorado Plateau is among the five most biologically diverse regions of the 110 defined ecoregions in North America, and it is likely the most linguistically and agriculturally diverse region on the continent. This "state of the ecoregion" report therefore models an approach for assessing not merely biological and cultural diversity and the interactions between them, but the unique assets of an ecoregion as well. The most unique assets of this ecoregion's biocultural diversity are its endemic species, native language isolates, distinctive agricultural and wildlands management practices, and incomparable traditional ecological knowledge. In particular, we argue that this ecoregion has sustained high levels of biological diversity and endemism relative to the rest of North America because of the heterogeneity of culturally based land uses, traditional practices of vegetation management, and informal protection of traditional cultural properties.

Given that this ecoregion is relatively arid, we place strong emphasis on how water management by various resident cultures has affected biological diversity. Because land and water management strategies, values, and management "scripts" are linguistically encoded in the many indigenous and non-indigenous languages extant in the region, the maintenance of these diverse languages is deemed essential to maintaining this diversity, as is tribal sovereignty and multicultural conservation collaboration. We recommend great investment in multicultural collaborations that can result in ecoregional conservation planning to include corridors between protected areas both on and off the Indian reservations, lands which comprise nearly a third of all land on the Colorado Plateau. This assessment explores the biocultural diversity on the Colorado Plateau by addressing three themes: (1) its geology, climate, geography, and hydrology; (2) the biological adaptations to these heterogeneous physical conditions that have led to adaptive radiation of species, varieties, and populations; and (3) the cultural diversity of the traditional ecological knowledge, values, and practices that have sustained it.

These three themes are explored in the light of four primary concerns: (1) known historic levels of diversity and the current status of surviving (extant) elements of diversity; (2) contemporary ecological and economic trends threatening this extant diversity; (3) plausible scenarios for losing or retaining the ecoregion's uniqueness; and (4) proposed options to safeguard and restore ecoregional diversity and uniqueness.

This assessment also demonstrates means by which multicultural collaborations can be promoted, incorporating the diverse voices of the region's residents into a "state of the ecoregion" report that can both inform and inspire future actions.

ACKNOWLEDGMENTS

We thank the many individuals who have been willingly interviewed as part of this assessment, and hope that their unique voices come through loud and clear, particularly those who participated with us in the two workshops that were held in the lovely retreat setting of the Arboretum of Flagstaff, where many of the recommendations offered in the following chapters were shaped (Steve Albert, Andy Bessler, Mae Burnett, Tom Carlson, Monica Castelo, Lisa Classen, Chris Coder, Marcelle Coder, Sunny Dooley, Zarina Estrada, Anthea Fallen-Bailey, Steve Gatewood, Greg Glassco, Julia Gold, Dave Harmon, Preston Hardison, Al Henderson, Daniel Higgins, Jennifer Holmes, Kristin Huisinga, Amanda Johnson, Ben Jones, Eljean Joshvema, Brett Kencairn, Leigh Kuwanwisiwma, Marvin Lalo, Louise Lockard, Steven Lomadafkie, Jonathon Long, Micah Loma'onvaya, Luisa Maffi, Vernon Masayesva, Jeremy McLain, Bianca Perla, Karen Sue Rolph, Jon Spence, Abe Springer, Ferrell Secakuku, Jennifer Sowerwine, Miguel Vasquez, Janet Vorhees, Diane Vosick, Pauline Watchman, Edward Wemytewa, Ken Wilson, Ofelia Zepeda).

Many tribal representatives gave generously of their time and also provided tribal data, especially on the status of indigenous languages on the Colorado Plateau. They were interviewed in person, via electronic mail, or by telephone by Patrick Pynes, Tony Joe, and David Seibert. A complete list of their names appears with Table 10 on pages 51–52.

The contributors also wish to thank Naima Taylor, David Seibert, and Patty West, as well as two anonymous reviewers for their constructive comments to an earlier version of this study.

CONTENTS

FOREWORD

The millennium—our millennium—arrived in the year 2000. Though millennial righteousness neither banished evil nor guaranteed peace, every encounter during that year seemed weighted with extra meaning.

In such a mood, one day that spring I talked for a spell with Larry Fletcher in Cannonville, Utah. The wind swept down on us from the west, whistling through the pines and firs shading the breaks at Bryce Canyon on the rim of the Paunsaugunt Plateau, then swirling through the pinyon-juniper forest on the benchlands dissected by the creeks whose slickrock narrows I'd been hiking with my family—Sheep Creek, Willis Creek, Bull Valley Gorge, all within the new Grand Staircase–Escalante National Monument—the wind warming, descending, cutting right on through the dusty badlands surrounding the Fletcher place on the outskirts of the little Mormon village on the banks of the Paria River.

Folks in town had referred me to the Fletchers: Larry, his brother Alma, and Alma's wife, Anita. After I had rattled our bike rack to pieces on the back roads of the Colorado Plateau, I needed a place to store our family's bicycles until I could return for them with a repaired rack—and the Fletchers had an empty shed in town. Larry was gracious, accommodating, and neighborly. His face was as wind-chafed as the barnwood of his outbuildings. And then he cocked his head and asked me a question. He wanted to know if I happened to be a member of the Sierra Club.

It seemed he had never met such a character. Larry had grown up in Cannonville. He imagined every Sierra Club member, every environmentalist, as his enemy—a faceless Other from New York, who had never been to his town or to his home territory, much less his driveway. He was convinced that these people hated him and wanted to destroy his way of life.

Larry looked at me with wonder when I admitted my identity. I was his enemy, but I didn't fit his caricature. Sure, we had plenty of differences. But we could talk without fisticuffs. I knew a few of the places that he knew, even a few of the same people. I valued the multi-generational commitment to the Colorado Plateau that Larry and Alma and their forebears had shown. Their determination to live here—evolving toward endemism—is a miracle of stability and identity in a changing world.

Larry would be intrigued with many of the facts in this report. He would like hearing that the Hopi language has names for at least 15 folk varieties of corn. He would approve of one of the report's conclusions: that the next century of regional conservation should not try to attract European tourists or literary naturalists but rather should aim at creating lasting benefits to the quality of life for the people who live here. Such permanent residents include the Mormon and Hispanic farmers and ranchers, but begin with the native people, who manage nearly a third of the plateau in tribal reservations.

I also believed that Larry wasn't the owner of all of that public land that surrounded his home, though he knew some of its nooks and crannies better than anyone. We hold this land in trust, together—as citizens, as a species. And so, as we begin our next millennium, we share the responsibility both to safeguard and restore the Colorado Plateau and to learn from each other.

This report concisely summarizes those responsibilities. These authors give equal weight to biodiversity and cultural diversity, and not just in a glancing paragraph of acknowledgment. They document the biocultural diversity of the Colorado Plateau with real data that justify the "unique" in the title. Refreshingly, they recognize that the future will be determined not just by federal land managers and scientists, but by tribal governments, by indigenous healers and farmers, and by Larry Fletcher, too. Interconnection, collaboration, and community are the keys to the future.

With its scenery and rivers and stunning light, we are used to superlatives applied to the Colorado Plateau. But now, in this "ecoregional assessment," we learn startling new superlatives. Of 110 defined North American ecoregions, it ranks in the top three for endemic species and in the top five for biodiversity—an unrivaled sanctuary for indigenous ethnolinguistic and agricultural diversity.

This is the richest living laboratory in the Americas north of the Tropic of Cancer for assessing the influences of diverse cultures on a biologically diverse landscape. But it's vulnerable, terribly vulnerable—to change, to growth, to land fragmentation. These superlatives apply to a place whose very name remains poorly known. These words, Colorado Plateau, do not yet ring with the clarion call struck by the two other North American places of similar rank, the Everglades and the Great Smokies. Perhaps this is one reason why the threats are numerous, relentless, local, regional, and global.

I've been writing about the Colorado Plateau for nearly 30 years. I've hit on a few phrases to keep with me, a rosary to repeat to myself as I continue to search for its core. Once, in trying to pinpoint the essence of its wildness, I called the plateau "the bright edge"—after a quote from Willa Cather, in which she perfectly articulated that "lightness, that dry aromatic odor" peculiar to the "air of new countries" that survives today only on the shining, wild, bright edge of the world. A few years later, I prayed, along with the Zuni firekeeper at winter solstice, that all of us on the plateau "be blessed with light," that we might be graced to live with the solace and exhilaration that this intensely colored landscape brings to us.

If the Southwest is Indian Country, the Colorado Plateau is its heart. On maps, the plateau is even shaped like a heart—a slightly cockeyed heart, cranked clockwise with the spin of geologic time, so that the cleft nestles close to Mesa Verde and the apex lies just about where the Colorado River leaves the Grand Canyon and loops away into the desert. The entire region is a single enormous sacred landscape: The Colorado Plateau is a prayer turned to stone.

This report proves what I've been suspecting: We can no longer depend on isolation to preserve diversity; we need to manage and plan and trade ideas. We need hybridization and convergence—while we celebrate uniqueness. We must pray here together—out on the slickrock, back in the alcoves, down on the river.

Scientists, indigenous people, wilderness zealots, federal land managers, and Larry Fletcher look out over the Colorado Plateau and share an astonishment, an awareness, a love, an urge to fight for the land. We will need to act together to accomplish what we all want—preservation of unique biocultural diversity. Tribal elders will need the skills of trained educators. Scientists will need to acknowledge their botanist counterparts, the native farmers. Ranchers will need to stand shoulder to shoulder with range managers and environmental activists and tribal religious leaders at springs and, together, agree to the compromises that will prevent their loss.

With its vital heart of indigenous culture, its low urban population, its extravagant diversity, its history of careful science, and its vast congregation of pilgrims and parishioners, the Colorado Plateau has the chance to balance the conflicting needs of all of its constituencies. This is not just a living laboratory for science, but a testing ground—a millennial test, after all—for our future as a community of human beings.

STEPHEN TRIMBLE

PREFACE

This report is the fruit of a collaboration established in 2001 between Terralingua: Partnerships for Linguistic and Biological Diversity, the Center for Sustainable Environments (CSE) at Northern Arizona University (NAU), and the Grand Canyon Wildlands Council (GCWC). Terralingua is an international non-governmental organization, founded in 1996, that is devoted to fostering research on the links between biological, cultural, and linguistic diversity worldwide and to promoting integrative policy and applied work in this biocultural domain (Maffi et al. 1999; WWF International and Terralingua 2000; Maffi 2001; Maffi et al. forthcoming; www.terralingua.org). Partnerships with like-minded organizations and institutions, such as the CSE and GCWC, are integral to Terralingua's work.

As a part of its program, in 2001 Terralingua initiated a project that goes by the name of the Global Biocultural Diversity Assessment (GBCDA), with initial funding from the Ford Foundation. The GBCDA, intended to be a multiyear project, aims to provide an integrated assessment of global biological and cultural (biocultural) diversity. The project thus proposes to go beyond previous assessments of global biodiversity, such as the United Nations Environment Programme (UNEP) Global Biodiversity Assessment (Heywood 1995), which did not develop an in-depth analysis of the complex relationships and mutual influences between humans and the environment. In particular, drawing inspiration from the later UNEP companion volume *Cultural and Spiritual Values of Biodiversity* (Posey 1999), the GBCDA is meant to turn the spotlight on the value of traditional ecological knowledge (knowledge, use, and management of the environment by indigenous peoples, minorities, and other local communities) for conservation and restoration of ecosystems, and on the role of local languages in the maintenance and transmission of this knowledge.

The GBCDA has three main components: a set of GIS (Geographic Information Systems) analyses of global correlations between biological and linguistic-cultural diversity; an index of biocultural diversity to assess and monitor global trends in biocultural diversity; and a series of case studies on the links between biological and linguistic-cultural diversity at subglobal scales (from regional to local) and on concrete action for joint biocultural diversity conservation and restoration.

The project is an interdisciplinary effort drawing from the natural and social sciences, both theoretical and applied, and calling for collaboration and partnerships with a large number of institutions, organizations, researchers, and practitioners from around the world. The primary uses of its results will be in planning and implementing biocultural diversity conservation and identifying priority areas and appropriate avenues for action. Researchers and practitioners working on biocultural conservation in indigenous and traditional communities around the world, including members of such communities themselves, should benefit from the global database and from the integrated perspective it will provide, as well as from the conceptual and methodological tools offered by the index of biocultural diversity and the case studies.

The GBCDA will also be geared to guiding the development of appropriate policies, both national and international. To this aim, Terralingua is collaborating with various international organizations, institutions, and initiatives, including the United Nations-sponsored Millennium Ecosystem Assessment. It is also seeking to inject its perspectives into international processes such as the World Summit on Sustainable Development, to be held in Johannesburg, South Africa in the fall of 2002, where some of the GBCDA results are to be presented.

This report on biocultural diversity on the Colorado Plateau is a pilot project conducted by researchers at NAU's Center for Sustainable Environments, the Grand Canyon Wildlands Council, and their collaborators within the region, in partnership with Terralingua. The report was planned within the framework of the GBCDA, with funding from Terralingua's Ford Foundation grant and other donors assisting the CSE and GCWC. The goal was to bring together and synthesize for the first time a wide range of data on the natural and cultural makeup of the Colorado Plateau, bringing out the close relationships, both historical and present-day, between humans and the environment in this North American ecoregion. The additional purpose of this synthesis was to provide a preliminary assessment of the current state of the plateau's geographic, hydrological, biological, ethnolinguistic, and agricultural diversity; to identify trends that this diversity is undergoing and some of the main threats that are affecting it; to envision future prospects if such trends and threats continue; and to outline suitable response options to counteract and if possible reverse negative trends. In taking this approach, the project also addresses some of the main concerns of major environmental initiatives and processes such as the Millennium Ecosystem Assessment and the World Summit on Sustainable Development, in terms of integrated ecosystem assessments and concrete action to halt ecosystem degradation and foster sustainable human-environment interactions.

The reason for choosing the Colorado Plateau for this pilot project can easily be gleaned from even a cursory reading of this report. The uniqueness of the plateau in both its biogeographic and ethnolinguistic composition jumps at the reader from the following pages. As the report indicates, the Colorado Plateau is one of the five most biologically diverse among North America's 110 ecoregions, and it is the most diverse region of this subcontinent ethnolinguistically and agriculturally. It is also especially notable for its high levels of endemism, both biological and linguistic. It thus represents an ideal laboratory for the study of biocultural diversity in this part of the world. At the same time, this exceptionally diverse region is increasingly under threat from many economic, social, and political forces, both local and global, lending urgency to the assessment of the status and prospects of its natural and cultural resources and to devising innovative action to protect and perpetuate its biocultural diversity.

The significance of the Colorado Plateau from a biocultural perspective is enhanced by the persistent vitality of a large number of Native American communities (American Indian tribes) in this region. Their traditional ecological knowledge, encoded in their languages, has historically guided their use and management of, as well as respect for, these lands. Their land-related knowledge and values remain remarkably intact today, but are undergoing rapid change. As the report underscores, this circumstance makes it crucial to recognize First Nations as key stakeholders and partners in conservation and restoration projects, and to consider the strengthening of their cultural and linguistic heritage as an intrinsic component of any such effort.

The report's value lies in its bringing together this whole range of issues from a number of perspectives, expressed in a variety of voices, for the primary benefit of stakeholders, policy makers, and the general public. Above all, the report should stimulate a great deal of new research on any and all facets of this complex and fascinating region and on their interrelations. Each of these facets calls for additional investigation, and could be the object of several theses, dissertations, and research projects. If "living as if biocultural diversity matters" is the goal, as the authors of this report advocate in their conclusions, the Colorado Plateau is an ideal place to start.

LUISA MAFFI
Terralingua: Partnerships for Linguistic and Biological Diversity

SAFEGUARDING THE UNIQUENESS OF THE COLORADO PLATEAU
An Ecoregional Assessment of Biocultural Diversity

The Colorado Plateau: An Orientation and Invocation

Gary Paul Nabhan

If there is an area of North American wildlands that is easily recognized by people around the world, it is the Colorado Plateau, but by image, not by name. Citizens of a hundred nations circling this planet can see color photos of the Grand Canyon, Delicate Arch, Monument Valley, or the cliff dwellings at Canyon de Chelly, and immediately acknowledge that these places occur in the red-rock country of the American West, where there are cowboys and Indians who still depend upon the bounty of the land for their physical and spiritual sustenance. And yet, few of them would call this landscape the Colorado Plateau. If they could name the region at all, they might simply call it Slickrock Country, the Painted Desert, or the Heart of the Old West. They may know it as the place where Butch Cassidy and the Sundance Kid jumped off a cliff into a river, where Thelma and Louise drove their car off a mesa top, or where Navajo detectives chase skinwalkers in the novels of Tony Hillerman. Few of us know it for all of its resplendent natural and cultural diversity.

Even most of those who have traveled across this region seldom think of its 526,000 sq km as one cohesive entity (Figure 1), for it is unspeakably diverse in its rock types, vegetative cover, and cultural occupants. Flanked on one side by the Rocky Mountains, and on the other by the low-lying Great Basin reaching clear to the Sierra Nevada, the 326,000 sq km core basin of the Colorado Plateau is drained by the Colorado River. The plateau extends from 34° to 43.5° N latitude, and from 105° to 114° W, encompassing eastern Utah, western Colorado, northeastern Arizona, northwestern New Mexico, and marginally, portions of southwestern Wyoming. Nevertheless, it appears to many as an amalgam of anomalies:

1. A semi-desert formed by rain shadows, with some of the highest densities of freshwater springs anywhere on the continent.

2. An altitudinal ladder from some of the oldest exposed rock on the continent to alpine tundra on island-like peaks reaching 4600 m in elevation.

3. A mosaic of cultural enclaves including hunting-gathering, farming, and herding people with continuous occupancy of their homelands for millennia, juxtaposed with recent arrivals who represent one of the largest human migrations of the last century—that of European, Asian, and Afro-American families to the "Sunbelt."

4. An astonishing diversity of languages still spoken side-by-side with one another, with speakers of one endangered dialect of Tewa surrounded by Hopi speakers, who are surrounded by Navajo speakers (the largest cohesive native language community on the continent), surrounded by English and Spanish speakers.

The Colorado Plateau has not been immune to the forces of demographic change, globalization, and homogenization that have affected nearly every other region of the world. Its human population has grown sevenfold since 1900, with accelerated rates of in-migration occurring since 1970. The population of the 31 counties with more than 50 percent of their areas within the Colorado Plateau boundaries was recorded by the U.S. Census Bureau as 2,225,573 in

Figure 1. The "core" of the Colorado Plateau (not including its tributary watersheds), courtesy of the CPLUHNA program.

the year 2000, with only 40 percent of its residents defining themselves as single-race Caucasians. In other words, 60 percent of the region's people define themselves as Native American, Hispanic or Latino, African American, or as people with two or more races in their ancestry. Today, the major population centers on the Colorado Plateau are Rock Springs, Wyoming; Grand Junction, Montrose, Cortez, and Durango in Colorado; Price, Moab, Kanab, and St. George in Utah; Farmington, New Mexico; and Page, Tuba City, Kingman, and Flagstaff in Arizona. Although the residential human population density of the plateau barely exceeds 2.5 persons per square kilometer, tens of millions more come to the region each year for recreational purposes.

What the human population lacks in density compared to other ecoregions in North America, it makes up for in antiquity. Oraibi, a village on the Hopi Indian Reservation, and Acoma Pueblo are reputed to be the longest continuously inhabited settlements in North America, and Paleoindian sites on the plateau suggest human influences on the faunal composition of the region dating back at least 12,500 years.

The region has been vulnerable to boom-and-bust mining schemes that have left its residents suffering from exposure to uranium and other environmental health risks. Decades of unregulated grazing practices have left rangelands impoverished, and have affected forest health as well. The more recent economic dependence on tourism and retirement communities has not always created more financial stability for long-term residents, but has aggravated issues of inequity and environmental injustice. The region is home to roughly twice the national average of people living below the poverty level, and per capita income on the Colorado Plateau has remained at a level that is one-third below the national average. The region will continue to have poverty and poor education as long as per capita educational spending in northern Arizona and northern New Mexico remains among the lowest in the nation.

And so, the following chapters highlight the many unique natural and cultural assets of the Colorado Plateau, without glossing over the very real threats to their continued existence. The authors explore the current barriers to adequate stewardship of the region's special qualities and resources, but also highlight success stories told by local community members who have found ways to safeguard what is most precious to them. It is not possible to provide equal coverage to all issues and topics in such a state-of-the-region report, but the authors have made every attempt possible to demonstrate how issues of biodiversity and cultural diversity are inevitably integrated at every imaginable scale of life within this region.

Assessing Levels of Biocultural Diversity on the Colorado Plateau in Relation to Other Regions

Gary Paul Nabhan, Patrick Pynes, and Tony Joe

> The canyon country of southern Utah and northern Arizona—the Colorado Plateau—is something strange, marvelous, full of wonders. As far as I know, there is no other region on earth much like it, or even remotely like it. Nowhere else have we had this lucky combination of vast sedimentary rock formations exposed to a desert climate, a great plateau carved by major rivers—the Green, the San Juan, the Colorado—into such a surreal land of form and color.
>
> —Edward Abbey (1977)

The Colorado Plateau is one of 110 defined ecoregions in North America that have been compared for their levels of biodiversity, the current degree of endangerment of species and biotic communities, and the relative strength of conservation measures already in place (Ricketts et al. 1999a). Covering watersheds encompassing approximately 939,000 sq km (9.4 million acres), the Colorado Plateau shrublands are situated between these other ecoregions: Great Basin shrub steppe and Wasatch-Uintah montane forests to the west, the Wyoming Basin shrub steppe to the north, the Arizona-Mogollon montane forests to the south, and the Colorado Rocky montane forests to the east. In addition, its biota continues to be influenced by the close proximity of the Sonoran, Mohave, and Chihuahuan Deserts, and at higher elevations, the semi-arid grasslands and woodlands of the Intermountain West.

The plateau exposes a greater diversity of geological strata and soils for erosion, or colonization by plants, than any other region in the Americas. This factor, combined with its broad elevational amplitude and highly dissected terrain, makes the Colorado Plateau an ideal setting for adaptive radiation and speciation. Compared to the other ecoregions in North America, the plateau harbors a remarkably high diversity of plants, butterflies, tiger beetles, and mammals, ranking in the top five ecoregions for species richness in these taxonomic groups (Ricketts et al. 1999a, 1999b).

Most interest in formally protecting the landscape integrity extant in the region was actually initiated long before our belated recognition that the Colorado Plateau was so biologically diverse. Nevertheless, because about 8.3 percent of the land area of the plateau is already protected by federal agencies, some global and national assessments do not consider the plateau's biodiversity to be as unique or as gravely imperiled as that in other regions of North America. For instance, the Nature Conservancy (Stein et al. 2000) ranked it only within the third tier of hotspots of imperiled biodiversity. However, another of their indicators of threatened diversity—rarity-weighted species richness—ranked the Colorado Plateau as being more biologically significant and less well known than it deserved to be (Stein et al. 2000). In other words, the plateau harbors many biological rarities whose vulnerability to threats has not yet been adequately assessed.

One flaw in most previous biological comparisons of the Colorado Plateau with other ecoregions is that species richness (the total number of species) is used as the predominant (and in some cases, the only) indicator of global biological significance. In this assessment, we place greater emphasis on biological and cultural uniqueness; we are skeptical that species richness of different ecoregions is a sufficient means to compare their ultimate conservation value. Key components of biocultural diversity in any region are its narrowly distributed endemic species, its unique biotic communities, its mosaics of communities juxtaposed in the same landscape, and the distinctive cultural influences on that landscape, particularly those emanating from indigenous cultures whose native languages are spoken nowhere else.

When these additional factors are taken into account, the Colorado Plateau must surely be ranked with the Everglades of southern Florida and the Great Smokies of the southern Appalachians as being among the three most bioculturally unique and diverse ecoregions in North America. It is already considered to be among the top three ecoregions in continental North America for total number of endemic species (Ricketts et al. 1999b), with the highest rate of endemism among vascular plants (Kartesz and Farsted 1999). Armstrong (2002) has determined that 23.6 percent of the mammals and 36 percent of the rodents on the plateau exhibit endemism at the levels of species or subspecies (Appendix 1). These are but two of many indicators that the Colorado Plateau ranks high among all North American ecoregions with regard to its biological uniqueness.

Few biologists in this region or elsewhere have fully reckoned with any ecoregion's biocultural diversity. That is to say, ecological assessments seldom focus on the inextricable links between biological and ethnolinguistic diversity, which through traditional ecological knowledge, values, and practices fosters a unique set of land use mosaics that influence biodiversity at the landscape level (Maffi 2001). Although there are now sophisticated ecoregional projects to assess biological diversity in the Great Smokies, Costa Rica, Greece, and elsewhere, there is little recognition of the influences that culture has had on this diversity, or appreciation of the indigenous knowledge of these plants and animals that is encoded linguistically in native languages.

> At Canyon de Chelly, floral and faunal diversity was intertwined with the cultural survival of the Anasazi, and the loss of biodiversity is probably linked to the Anasazi abandonment of the area.
>
> —Teresa N. Stoker (1990)

The Colorado Plateau is perhaps the best living laboratory in North America for assessing the interactions of diverse cultures on a biologically diverse landscape; although the Great Smokies have Cherokee influences and several Seminole groups reside on the edges of the Everglades, no other region of North America is home to so many cultures with such long tenure of interacting with land, water, plants, and animals. More Native American tongues survive here than in all other regions in the United States combined, including some 22 indigenous languages or dialects as well as English, Spanish, and Basque. Some of them, such as Zuni, are language isolates or endemic tongues of the Colorado Plateau with no close relative ever identified, whereas the Keresan languages spoken at Laguna and Acoma Pueblos and the Kiowa-Tanoan language spoken at Jemez Pueblo are relatively distinct from others in their language families. Not a single conservation plan, however, takes into account both the cultural diversity and biological diversity of the region—as if the historic and geographic relationships between these two factors are not relevant. As a result, the management of cultural resources has typically been done by different sets of specialists, sometimes involving Native American residents in the former, but regularly ignoring their traditional knowledge in the latter.

The historic failure of conservation biologists and environmentalists to engage Native American communities in collaborative work based on shared goals is short-sighted, because the long-term residents of these reservations know quite a lot about the history of the local flora and fauna. Although their traditional ecological knowledge (Berkes 2000) remains relatively intact, it has hardly been taken into account and valued by land managers, who often consider it to be composed of the anecdotal or superstitious recollections of scientifically untrained oldtimers.

Even the potential contribution of contemporary Native Americans in managing lands of the Colorado Plateau has been chronically underestimated. The Grand Canyon Trust considers "Indian country" to comprise just a quarter of the Colorado Plateau's 526,000 sq km area (Hecox and Ack 1996; Wilkinson 1999), but our calculations more accurately consider 31–32 percent of the plateau to be currently managed by sovereign Native American nations (Figure 2). By virtue of this fact alone, it would be presumptuous if not impossible to develop an ecoregional conservation plan or biodiversity assessment that did not take into account Native American stewardship of nearly a third of the land on the Colorado Plateau (Nabhan 2000a, 2000b).

Biodiversity Conservation in a Culturally Diverse Region

The cross-cultural issues to be addressed here are not peculiar to the Colorado Plateau, for they are to some extent echoed in other regions of North America and on other continents as well. Currently, Indian reservations in the United States cover 22,257 sq km (55 million acres), more land than is managed by the National Park Service; on other continents indigenous people also inhabit the last vast tracts of intact wildlands in their nation-states. Reservation trust lands cover nearly four times the area that national parks, monuments, preserves, conservation areas, wilderness areas, and wildlife refuges cover on the Colorado Plateau. Nevertheless, the lands managed by Native American stewards on the Colorado Plateau, about 16,630 sq km, have yet to receive much investment from federal or private sources for the inventory of their biodiversity, monitoring and recovery of their rare species, or local capacity-building in environmental protection—relative to the considerable support given to those working on adjacent federally protected lands.

> Key in [conserving biologically diverse places] will be understanding those places well enough to know what form protection must take. At the same time, however, we need to revisit the notion of place ... We are challenged with considering not only what places are most important to protect but also what is needed to ensure the long-term survival of their species and ecological communities.
>
> —Bruce Stein (Stein et al. 2000)

Non-indigenous conservation biologists may feel ill-equipped to deal with the cultural and legal (sovereignty) issues regarding plants and animals restricted to tribal lands, but they can no longer ignore the fact that the only means to sustain these species is by providing indigenous land managers with the resources needed to protect or recover these rare populations and their habitats. And yet, while indigenous people are often considered to be "land-rich" and "knowledge-rich," they collectively remain poor in the kinds of resources usually deemed essential for successful conservation work. The counties on the Colorado Plateau in which Native Americans are the predominant residents register much higher than average levels of poverty. In both obvious and subtle ways, this endemic poverty has limited members of the region's many diverse cultures from becoming full participants in conservation collaborations and other discussions about the future of their homelands.

Integrating Conservation Efforts and Biocultural Diversity

Over the last decade, tremendous effort has gone into defining, locating, and rapidly assessing the species richness of "hotspots of biodiversity" (Mittermeier et al. 1998; Ricketts et al. 1999). By some accounts, these hotspots not only capture a large proportion of the planet's biodiversity, they also define one scenario for conservation investment—that of outright purchase of wildlands and

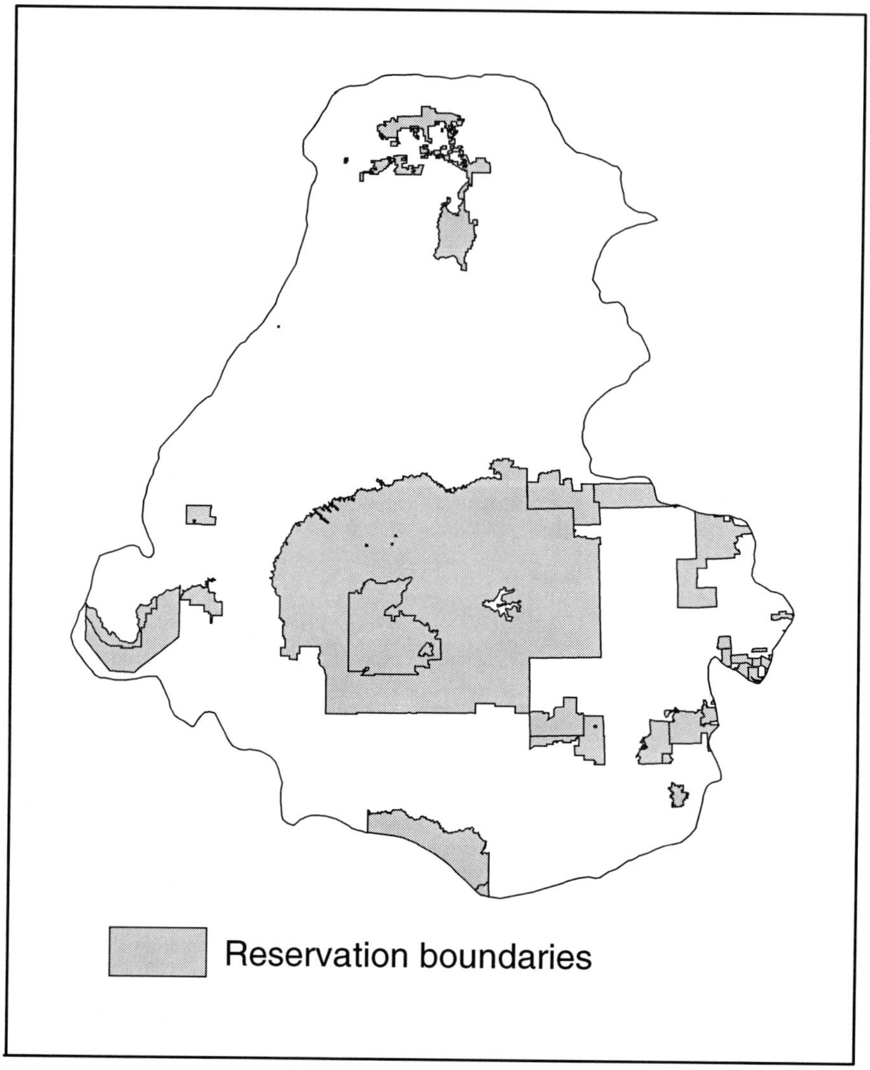

Figure 2. Nearly a third of the Colorado Plateau lies within reservation boundaries.

development of protective infrastructures to better protect vulnerable resources (Mittermeier et al. 1998).

By the mid-1990s, prominent ethnobiologists, anthropologists, linguists, and biogeographers had brought attention to the fact that the 10–20 hotspots richest in biodiversity were also extremely rich in cultural diversity, defined in terms of the richness of extant indigenous languages (Maffi 2001). Clearly, cultural land use patterns on the Colorado Plateau make evident a pattern seen on other continents as well: it is unlikely that the extant cultures dwelling on the land in these hotspot ecoregions are willing or able to allow their remaining territory to be designated as areas for the protection of wildlife, from which they themselves will be excluded. In fact, the "land purchase in hotspots" strategy, even if it were to become an acceptable scenario to indigenous communities, would in many places turn out to be counterproductive. It appears that where hotspots of biodiversity or endemism are inhabited by a diversity of cultures, those cultures already encode traditional ecological knowledge about species in their native languages, which has tremendous potential for helping to conserve biodiversity (Nabhan 2000b). If this is true, several alternative scenarios must be considered:

1. Government and non-government organizations need not always purchase the lands within these hotspots for biodiversity protection, because many of the biologically diverse areas are already "common lands" held in trust by these cultural communities, such that they cannot be purchased, traded, or condemned. New efforts to conserve biodiversity elsewhere in the world need not make the same mistake the U.S. Park Service historically made on the Colorado Plateau, by stealing, condemning, or finagling the rights to species-rich lands held in trust for future generations of indigenous people (Burnham 2000). Instead, investment in capacity-building and jobs for traditional stewards of the land may be preferable.

2. It is likely that proponents of biodiversity conservation will forge stronger, more effective collaborations with indigenous residents in biodiversity hotspots if they fully listen to and heed indigenous concerns about sovereignty, cultural property rights, and secrets associated with esoteric beliefs and ceremonies. Conservationists should attempt to collaborate on community-based projects that foster the continued oral transmission of traditional ecological knowledge, the persistence of indigenous languages, traditional subsistence or ceremonial practices, and the generation of health benefits or income for a broad cross section of community members. Although such community-based conservation efforts are already given lip service from governments and NGOs alike, there remains a disproportionate investment in top-down conservation strategies, and only scattered investment in community-based, bottom-up strategies for maintaining biodiversity (Shand 1997).

3. Given that the traditional ecological knowledge about biodiversity itself is being diminished, the fragile relationship between the two needs to be more tangibly explored. Many biogeographers and linguists are already involved in efforts to understand global patterns of biocultural diversity (Maffi 2001), but it is hoped that our pilot project to safeguard the uniqueness of the Colorado Plateau will encourage other community-based practitioners to experiment with more practical means of integrating indigenous knowledge into collaborative efforts to conserve other hotspots of diversity and endemism.

Eliciting Perceptions of Biocultural Diversity on the Colorado Plateau: A Methodology for Defining Threats and Response Options

Thomas D. Sisk

> The People of the Colorado Plateau today are the product of more than five centuries of interaction and change. Nowhere else in the United States do so many different peoples converge. Nowhere else in the United States is the history of the land and its peoples so closely tied together.
>
> —Kenneth Brown (1995)

Efforts to safeguard biocultural diversity on the Colorado Plateau presuppose that biological and cultural diversity is valued, that these values are threatened, and that people can do something about it. Without the appreciation of the richness of nature and its human traditions, the idea of conservation is superfluous. Without recognition of threat and a sense of loss, human attentions will focus elsewhere.

The widespread, public commitment to cultural and biological conservation on the Colorado Plateau—from community groups to national environmental organizations, from tribal governments to federal land management agencies—demonstrates a general consensus that the uniqueness of the plateau is threatened, and that these unique resources are eroding. But what characteristics of the region are at risk, and what is the nature of the threat? Strip mines and belching smokestacks come to mind, but what about urban sprawl and global warming? Global trade and television? All of these have been suggested as threats to biocultural diversity. If we accept them all, and add others to an ever-lengthening list, which of these threats are most important and merit our immediate attention? How might we respond?

These questions are multilayered and complex, and how we answer them forms a critical step in the design of effective conservation actions. Here, we take a first look at these questions, while developing a framework for analysis that serves conservation biologists, language revitalization educators, resource managers, activists, and policy makers. We identify two key categories of conservation focus—resources and potential threats—and we examine the overlap between their high-priority elements. We suggest that conservation actions should be focused where unique and important resources face impending threats. When these focal areas are identified, appropriate actions can be contemplated, designed, and undertaken.

The analytical approach developed and illustrated here is employed with greater detail and differing emphasis in subsequent chapters. Throughout, we strive to honor the diversity of the region and embrace its awe-inspiring complexity, while providing a conceptual framework that brings clarity to our understanding of current conditions, trends, and opportunities for protecting the unique biocultural resources that characterize the Colorado Plateau.

Approach and Methods

Our objectives in this project are ambitious. As a starting point, we turned to practitioners of biocultural conservation, the hundreds of professionals who know the region well and who work with these issues day after day. These professionals have collectively contributed more than a thousand "person-years" of observation and research to the resources on the plateau. Rather than identifying and elevating any single particular view, opinion, or tradition, we began this study with an attempt to incorporate the diverse perspectives and detailed knowledge of these practitioners into an inclusive treatment of the common concerns and shared vision for the region's future. Drawing on the practitioners' responses to an initial survey, as well as the additional comments they offered, we consider the physical, biological, and cultural resources thought to be at greatest risk. Next, we explore the concept of threat and examine how a heightened appreciation of contemporary threats can inform our conservation actions. Finally, we consider appropriate responses, asking how people might most effectively reverse the erosion of cultural and biological resources occurring across the Colorado Plateau.

Asking the People: The Survey

Working in consultation with Thomas Combrink, senior research specialist for Northern Arizona University's Arizona Hospitality Research and Resource Center, we created a 7-page survey, "Safeguarding the Unique Resources of the Colorado Plateau." A kind of extended questionnaire, the survey included both objective and narrative sections, and was distributed to approximately 400 individuals working on the plateau. These 400 people were selected from a longer list of natural and cultural resource professionals that we compiled from a variety of sources, including our own knowledge of people, organizations, and communities. The result was an extensive directory of people from a variety of ethnic, cultural, and professional backgrounds. Common among them, however, were years of on-the-ground experience and a personal commitment to natural and cultural resource issues on the Colorado Plateau. The vast majority of the people we surveyed are, or were, long-term residents of the plateau. They include state and federal agency personnel, university scholars, researchers, teachers, tribal environmental professionals, community activists, and many others whose primary occupations revolve around the management and conservation of biological and cultural resources on the Colorado Plateau.

Of the more than 400 surveys that we distributed, 90 (22%) were returned to the Center for Sustainable Environments, and of these 71 (18%) were sufficiently detailed to contribute to quantitative analysis (a relatively high rate of returns for mail-out surveys). The results were analyzed by Center for Sustainable Environments researchers working with the Arizona Hospitality Research and Resource Center, who compiled selected excerpts from the survey's narrative sections. Some respondents to the survey wrote long, moving passages in the process of answering several open-ended questions. Selected excerpts from these responses are scattered throughout the text.

The completed surveys provide several kinds of intriguing and useful information, particularly regarding the nature and categories of biocultural resources, the nature and importance of perceived threats, and the effectiveness of various approaches to conservation. These completed surveys were our initial guides into the complex arena of analyzing biocultural diversity on the Colorado Plateau; collectively, they form the core of this chapter and an important foundation for this report.

Naming Biocultural Resources

As a first step, we must identify what we value as biocultural resources. For many, this is a difficult and awkward task. The plants, animals, soils, and springs that characterize the region take on different identities and embody wildly varying values among the people of the Colorado Plateau.

For many, specifying definitions and quantifying values is antithetical to the integrated physical, biological, and cultural reality of ancient tradition. Yet most people realize that some such effort is needed to bring people together in a shared recognition of values and a common effort to protect them. In our survey, we asked people to rank the significance of eight categories of biocultural resources according to their contributions to the uniqueness of the Colorado Plateau ecoregion. In addition, each respondent was asked to identify the most significant or highly valued resources within each of these categories, generating a much more detailed perspective on unique resources. These biocultural resource categories (Table 1) were outlined in the following manner: (1) geohydrological features, (2) biotic communities, (3) traditional cultural properties (places), ceremonies, or art forms, (4) ecological processes and interactions, (5) endangered or threatened animals, (6) endangered or threatened plants, (7) distinct languages or dialects, and (8) native crop plants and minor livestock breeds.

Many Voices

From a diverse group of respondents came a diverse set of perspectives and values. These were conveyed with an intensity that is almost palpable, reflecting the importance placed on these resources by the people working close to the land. The length of the passages, the place-specific detail, and the blending of objective analysis with impassioned imagery characterize most of the completed surveys that we received. The following statements are edited excerpts from survey responses. Together, they highlight the biocultural values ranked highly by our respondents, while providing some idea of the breadth of perspective and depth of commitment to the plateau's unique biological and cultural heritage.

On Geohydrological Features—

> Water and waterways set the potential for plant, animal, and human populations on the Colorado Plateau.
>
> It is the character of water resources that makes possible the continuation of native lifeways in a manner that reflects the traditional heritage of many of the tribal peoples.
>
> Water is the lifeblood of this semi-desert region. Living communities are forever changed when we tinker with its flow. Perhaps it is arrogant to control water for human benefit without much regard for the rest of the ecosystem.
>
> Every waterway or spring is highly significant.

On Biotic Communities—

> Riparian areas on the plateau, being in essence rare, need to be protected from overgrazing and ORV [off-road vehicle] abuse. They are the arteries of plateau life and are threatened.
>
> The declining health of the ponderosa pine forests is troubling. These forests are at a real risk of near total destruction by catastrophic fire, with a subsequent collapse of the reproductive cycle.

On Traditional Cultural Properties, Ceremonies, and Art Forms—

> Historical artifacts are a window to the past, but human culture is plastic, inevitably mutable. Rock art provides an awareness of and a link to prehistoric people who, in their own way, altered the landscape. It provides the opportunity to extend our limited knowledge about the past.
>
> The use of native plants by the early people has been lost in some cases, and if possible needs to be revived. There is real knowledge and wealth in this for humans.

On Languages—

> The language is what makes us so special and distinct. When we lose it, we'll be just like the rest, and this feels scary.
>
> All indigenous languages, as well as New Mexican–Southern Colorado Spanish, are highly significant resources. The loss of native speakers since 1945 has been tremendous, threatening the continuation of much indigenous knowledge of the Colorado Plateau.
>
> It is the maintenance and continuation of the diversity of languages that makes this area unique.

Table 1. Ranking of perceived importance of biological and cultural resources on the Colorado Plateau, based on 71 responses to a questionnaire sent to more than 400 environmental professionals.

Biological and Cultural Resource	All Respondents (71)	Federal Agencies (19)	State Agencies (7)	Universities (21)	Reservations (2)	Environ. Orgs. (10)
Geohydrological features	1	1	1	1	5	1
Biotic communities	2	2	2	4	5	3
Trad. cultural properties, ceremonies, art forms	3	5	2	1	5	3
Ecological processes and interactions	4	3	2	5	8	2
Endangered or threatened animals	5	4	7	6	1	5
Endangered or threatened plants	6	5	5	7	3	6
Distinct languages and dialects	7	8	5	3	1	7
Native crops and minor livestock breeds	8	7	8	8	3	8

Fifty-nine of the 71 respondents were grouped into five categories, based on professional affiliations (12 did not self-identify). Ranks range from most important (1) to least important (8). Ties are given the same rank, and the next lowest rank is skipped; thus all ranks can be compared among categories of respondents. Additional resources identified by the respondents, and their perceived importance, are discussed in the text.

Common Perspectives

Across the range of responses, comments, and rankings emerged one clear perspective: at the root of the plateau's uniqueness flows fresh water. Almost all surveys placed the highest value on geohydrological features, with many articulating the linkages between the region's springs, seeps, rivers, and washes, and the biological resources and cultural traditions that also emerged as highly significant. It is impossible to overstate the role of water in sustaining the uniqueness of the Colorado Plateau. Almost a quarter of the respondents discussed the central importance of groundwater, seeps, springs, and rivers to biocultural conservation, with most environmental professionals extending their commentary to the vital role of riparian habitats associated with permanent or seasonal surface waters. Several respondents also discussed the "hanging gardens" clinging to cliff faces near springs and seeps and the integrity of the region's aquifers. In addition to riparian habitats, pinyon-juniper woodlands, ponderosa pine forests, and arid grasslands were mentioned as highly significant biotic communities. Ecological processes, including floods, fire, and the effects of global climate change on regional precipitation were discussed under the heading of ecological processes, along with migration for both terrestrial and aquatic organisms.

Perhaps it is surprising that endangered plants and animals ranked somewhat lower for most respondents, although declining native fish and other species, listed under the Endangered Species Act, and the Navajo Churro sheep were mentioned by multiple respondents. The latter was also mentioned frequently in the context of native crops and livestock breeds, along with Native American crop plants. Rock art was the most frequently identified traditional cultural property, although the lack of specificity provided in this grouping clearly reflects the small number of survey responses received from tribal environmental professionals, and this low response rate itself is undoubtedly related to the preference of native people to protect the identity of many sacred places and certain traditional knowledge. Distinct languages and dialects ranked lower, overall, than one might expect, given their centrality to the region's human communities and their role as, arguably, the strongest indicator of the region's cultural diversity. The rankings of these eight categories of biological and cultural resources are summarized in Table 1.

Striking congruence is seen in the top-ranked groupings, but this congruence can easily be overstated when all issues in the region are taken into account. The categories rated as most significant include mostly physical and biological resources, elements that are valued and understood by all people living and working on the plateau. Considerable differences of opinion emerge among the rankings for cultural resources, such as crop diversity, language diversity, and traditional cultural properties and ceremonies. The great diversity of cultural perspectives across the Colorado Plateau ensures that there will be widely divergent views of their importance, so that lower rankings for these issues are inevitable. More instructive, perhaps, is the variation in rankings for cultural resources among the five categories of self-identified professional affiliations (Table 1). Safeguarding native languages against threats is the top priority for some reservation-based environmental professionals, but the lowest priority for respondents from federal agencies. Traditional cultural properties, native crops, and livestock breeds span a similar range of rankings, but with less-intuitive patterns emerging among professional affiliations. These patterns may show as much about the limitations of surveys as they do about the priorities for biocultural conservation. Nevertheless, they do offer a starting point for gathering perspectives, assessing values, and charting a course for more in-depth inquiry and analysis. They also call attention to the interconnections between the identification of important resources and our perception of their vulnerability. Much of the variation in rankings shown in Table 1 may arise from differences in the respondents' exposure to cultural resources and traditions, and to their differing perceptions of the degree to which these biocultural resources are at risk.

What Is a Threat?

Threats are sources of injury, danger, or harm. They place valued resources at risk of degradation or loss. Threats to biocultural diversity can take many forms and emerge from many sources. They jeopardize indigenous cultures and native species, or alternately, benefit one at the expense of the other. Yet despite their net negative impacts on the land and on human traditions, threats are often hard to identify, define, and quantify (Czech et al. 2000). Their perception is subject to shared values about a resource that is threatened; without a common appreciation of what is placed at risk, there is no awareness of threat. Thus, the concept of threat itself is subject to cultural interpretations, and its meaning depends on knowledge of the land, its people, and their shared values.

Two examples illustrate the challenge of understanding threats to biocultural diversity. First, consider the damming of rivers and diversion of their flows for irrigation and the generation of energy. These actions can be seen as a risk to downstream villages and ecosystems that are denied flood control and other hydrological services that have been provided for millennia by the intact ecosystem. However, from the perspective of an upstream power user or distant urban dweller, the redirection of water provides a resource, and the "threat," if it is perceived at all, is identified with the failure of the dam and power delivery mechanisms, not their very existence. Of course, if we value the long-term provision of goods and services to human communities from healthy ecosystems, depletion of water resources is negative for all, but the point here is that different people may have strikingly different perceptions of "risk." For a second example, consider the region's arid grasslands. The past century saw a tremendous increase in livestock grazing as a source of livelihood on many Indian lands, and rural self-sufficiency and a community-based barter economy depend upon cattle, goats, and sheep in these regions. Yet overgrazing has devastated some areas and ushered in desertification, threatening native vegetation and soil productivity. Pinning down the threat in this case is a difficult search for an elusive idea.

How then can we identify and quantify threats to the biocultural diversity of the Colorado Plateau, without inserting geographically and culturally biased views? Is consensus possible? Perhaps it is not likely on any given issue, but we believe that a multifaceted approach to identifying threats can provide a more comprehensive and more helpful understanding of the trends that place biocultural values at risk. In some cases, values are in conflict and the balance between benefit and loss may remain unclear. In other cases, the relative importance of various threats may defy easy judgments. However, the shared desire for a future that safeguards the cultural and biological richness of the Colorado Plateau leads us to the task of examining the many advancing threats, sorting through them, and developing an integrated response that will help us into a future that retains and values the Colorado Plateau's diversity of people, ecosystems, and understandings.

> As we learn more about how natural ecosystems work, we are drawn inevitably to thinking on broader scales, both geographically and temporally. To be successful, though, we must take that grand vision and translate it into action at a particular place inhabited by real people. Ultimately all conservation is local.
>
> —Bruce Stein (Stein et al. 2000)

We begin by returning to our discussion of biocultural diversity from the previous chapter, and by framing the concept of "threat" at the scale of the human and ecological communities of the Colorado Plateau; that is, we take a biocultural view of threats to the biological and cultural integrity of these communities. To the extent that this view privileges the values and perspectives of plateau communities over those that derive "downstream" benefit from the plateau, so be it. As a discrete, unique, and globally important bioregion, one that supports the oldest continuously inhabited human communities in North America, the continent's greatest linguistic diversity, and high levels of endemism, the Colorado Plateau demands attention in its own right. Biocultural preservation must be the foundation of any planning efforts at national, continental, or global scales.

Three Dimensions of Threat

Threats come in various shapes and sizes. Among the broad range of stresses that currently place the plateau's biocultural diversity at risk, we examine three elements of threat, in order to sharpen perspectives. Drawing on the framework of rarity advanced by Rabinowitz and colleagues (1986), we consider the magnitude of the threat, its geographic extent, and its immediacy. By considering these three characteristics, we can identify several classes of threat, ranging from those that are local, manageable, and most likely to manifest in the future, to the most potent, widespread, current threats that demand immediate attention and coordinated responses. We ask three questions:

Magnitude—How great an impact will the threat have on biocultural values? Will it degrade the biological or cultural resource marginally, or threaten its very existence? We are concerned with threats that have modest, immediately apparent effects—such as the establishment of a single invasive weed in farmlands—but we also face threats of greater magnitude that are less obvious, such as the contamination of aquifers and the loss of traditional ecological knowledge.

Geographic extent—How expansive or localized is the perceived threat? Does it range across the entire plateau, like the effects of global climate change, or is it confined to a particular location or ecosystem, such as the impacts of a toxic spill? Large or small extent does not alone measure the importance of a threat, but it suggests much about the response required to address it.

Immediacy—Is the threat currently affecting biocultural values, or is it absent or insignificant at present, but likely to exert greater influence in the future? Climate change may be less important than overgrazing today, but its impacts may overwhelm any adjustment of grazing practices 30 years from now. Different responses are called for, depending on the immediacy of the threat.

Appreciating the Different Types of Threat

Humans have not been adept at anticipating threats to cultural and biological diversity. Historically, it has been very difficult to anticipate the effects of change. Most technological "advances" brought with them unappreciated problems, and early concerns often dissipated, whereas much more serious threats may initially have seemed benign. Changes in land use, economics, and languages have profoundly affected the uniqueness of the Colorado Plateau. Inmigration and settlement, development, education, and transportation have all brought powerful new forces that conveyed unanticipated benefits and costs. It is tempting to view all of this change, particularly the rapid changes that have been manifest since Euro-American colonization of the Colorado Plateau, as one amalgamated threat to the region's biocultural integrity. Satisfying as this may be on some levels, it only masks ongoing changes and continuing threats, and puts us in an even weaker position to combat the erosion of biocultural values. Instead, we attempt to disentangle various threads of change, and to examine potential threats. By taking a multifaceted view of perceived threats, we hope to be able to chart an effective set of responses to safeguard the uniqueness of the plateau.

When the magnitude, extent, and immediacy of threats are considered simultaneously, we see a diverse array of eight classes of threats. Here we briefly describe each class by providing examples.

1. Pressing problems requiring immediate, coordinated actions at all levels. These are the "crisis issues" that cannot be ignored but are difficult to solve. Regional decline in air quality is one example. Widespread overgrazing is another, as is the loss of native languages and traditional knowledge in the homogenizing wake of a global consumer culture.

2. Broad, emerging problems requiring integrated planning and actions. Global climate change is a prime example. Proliferation of road networks and appropriation of surface water are currently serious local threats that will soon threaten the entire plateau.

3. Pressing local issues requiring immediate action. Groundwater depletion and contamination threaten human communities scattered across the plateau. Although fairly isolated, these cases put communities at great immediate threat and require swift remediation.

4. Emerging problems that can be addressed effectively at the local level. Economic decline and educational stagnation in small communities threaten cultural and biological values by eliminating opportunities for residents and intensifying material needs that may lead to the further decline of local resources. Investment in communities and development of economic and educational opportunities can reverse rural decay and strengthen traditional values. While coordinated actions are needed, the focus should be on improvement at the scale of local communities.

5. Broad but manageable threats that require coordinated action. The loss of locally adapted varieties of crop plants and domestic animals is a threat to cultural and biological values that might be mitigated or reversed through concerted efforts to identify, propagate, and distribute traditional cultivars. Long-term success, however, also requires the retention and sharing of knowledge of traditional agroecosystems. Coordination of genetic research, propagation and breeding programs, and educational efforts is needed.

6. Emerging issues that can be addressed through integrated planning efforts. Rapidly increasing recreational use of the Colorado Plateau threatens sensitive plant communities and small-town economies via increased human presence and the establishment of the infrastructure to support it. Restriction of off-road vehicles and protection of vulnerable ecosystems and human communities through better recreational planning could decrease these threats while providing opportunities for the increasing numbers of people drawn to the natural beauty of the plateau.

7. Current problems that can be ameliorated through informed management actions. Tensions between private landowners, federal agencies, and tribes hinder integrated planning and management of natural and human communities. For example, enforcement of the Endangered Species Act should benefit all, but jurisdictional issues and misunderstanding often constrain decision making, leading to the implementation of inadequate plans or the ignoring of conservation imperatives. Improved intergovernmental mechanisms for addressing endangered species issues could foster plateau-based decision-making that will support effective on-the-ground management.

8. Perceived threats that can be avoided through informed decision making. The key to the future probably lies in dealing with this overlooked category of threat. Emerging problems that seem less important, local, and not yet critical are tomorrow's big issues. Left unanswered, they often grow in magnitude, extent, and immediacy. Current trends in suburban development, second homes for urban dwellers, and the resulting increases in property values may dramatically change demographics on the plateau, drastically changing the regional economy and marginalizing long-term residents and traditional communities. Long-range planning could encourage thoughtful integration of old and new, but history suggests that such foresight will be difficult to achieve. If these patterns continue, market forces will likely restructure the plateau's human communities, introducing novel threats to many biocultural resources.

What Are the Threats to the Colorado Plateau?

Specifics are everywhere. From increasing human populations to greater mechanized travel, mining, and energy generation, threats to the biocultural values of the Colorado Plateau are undeniably pervasive. Indigenous and traditional people agree that these threats are increasing in number and magnitude, but they disagree as to their identity and relative importance. Reductionist scientific approaches cannot assess the differing values that interweave cultural and ecological communities. Therefore, such approaches cannot produce an objective analysis. Long lists of threats provide limited understanding, just as lists of plant species provide limited information about their medicinal or ceremonial uses, or their roles in ecological communities.

Our survey initially addressed the issue of threats by asking the following question: "In your opinion, what are the most important threats to the distinct character and integrity of the Colorado Plateau?" This request for informed opinion generated a wide range of responses, described in part below. Respondents were then asked to rank a list of 15 perceived threats that we provided in the questionnaire. Listed in random order, these threats were construction and maintenance of roadways, urban and suburban growth and land fragmentation, demographic shifts affecting rural and urban ethnic mixes, influences of mass media and advertising, invasions of non-native plants and animals, influence of global free trade on local economies, conversion of wildlands to agriculture, poor grazing practices, poorly managed fuel loads and fire regimes, damming, depletion, or diversion of water, mining of coal, petroleum, natural gas, or minerals, logging or other extractive uses of plants, expansion and intensification of tourism and recreation, changes in land tenure and cultural values affected by migration from other regions, and devaluation of traditional land practices and related cultural values.

Although this list was clearly incomplete and perhaps unintentionally biased, it did provide a single list of commonly identified threats that could be ranked according to their relative importance. By examining the overall weights given in this list, and by comparing the responses of different subsets of respondents, we were able to identify common perceptions and values among natural and cultural resource professionals.

Many Voices

What are the most important threats to the character of the Colorado Plateau? In addition to the 15 threats listed on the questionnaire, individual responses identified cultural homogenization, global climate change, contentious water rights, lack of focus on biodiversity in management decisions, intolerance of other cultures and values, lack of adequate governmental representation, cultural traditions that are not ecologically sustainable, elevation of individual rights over the common good, political corruption, anti-government sentiment, noise pollution, light pollution, vandalism and the destruction of sacred sites, political pressure for privatization of federal lands, and lack of regard for human rights. Many respondents were eloquent in their descriptions of these threats. Each of the following quotations comes from a different respondent, describing a particular threat to the plateau:

> Threats come from numerous development efforts, which are driven by increasing populations. There are vast areas still free from the pressures that come with population increase, but each year more is touched by the human presence.

> With the increase in population has come an attitude of land, place, and cultural "experience" as a commodity. The reduction of diversity into a commodity has weakened the sense of this place as different.

> Observe Moab and surroundings for a weird, short-term example of unacceptable change. Although some are appalled, many thrive on the carnival atmosphere Moab now provides. In this case, development and progress might be more accurately termed exploitation.

> The globalization of residents, including Native Americans, so that few people have a culture that grew from and blends with the land ... [and] top-down government ... brings in more outside experts, homogenizes prescribed solutions, and relies even less on local indigenous stewardship.

> Policies [are] often driven by corporate entities [who] place profit at odds with practical, common-sense approaches to protecting ... resources.

> National Monuments being established without consultation with the native people who have claims to the areas as traditional cultural properties.

> Plants that [provide] food and medicine for cultural and religious ceremonies are ... becoming scarce. The atmosphere we breathe is also threatened by the uranium mines that have never been cleaned up.

> Encroachment of the dominant culture ... [and] lack of understanding about how the whole people/land/culture synergy works.

Common Perspectives

People's answers to open-ended questions about threats were broad and varied. The majority clearly focused upon the threats posed by expanding human populations, the erosion of traditional cultural values, and the introduction of external economic forces, or "globalization." Charles Wilkinson (1999) has called this unrelenting pressure on the Colorado Plateau the "Big Buildup," a process that has driven ever-increasing demands for the ecoregion's water and energy resources by the booming Southwest cities that have grown up around its periphery, especially since World War II. Many respondents connected population growth and economic effects in their narrative responses, but the task of ranking the list of 15 threats resulted in a less unified perspective on development pressures. These and all results from the threats rankings are summarized in Table 2.

When asked to rank the 15 threats on the survey form, "urban and suburban growth and land fragmentation" ranked first, whereas "influence of global free trade on local economies" ranked last. The fact that globalization ranked last as a perceived threat may be surprising, as many higher-ranking threats are driven by activities

Table 2. Ranking of perceived threats to biocultural values on the Colorado Plateau, based on 71 responses to a questionnaire sent to more than 400 environmental professionals.

Threat	All Respond. (71)	Federal Agencies (19)	State Agencies (7)	Universi-ties (21)	Reserva-tions (2)	Environ. Orgs. (10)
Urban and suburban growth and land fragmentation	1	1	3	1	1	1
Damming, depletion, or diversion of water	2	2	1	2	2	3
Poor grazing practices	3	4	2	3	4	2
Invasion of non-native plants and animals	4	3	4	8	9	9
Mining of coal, petroleum, natural gas, or minerals	5	6	6	4	3	5
Devaluation of traditional land practices & cultural values	6	13	11	9	5	10
Changes in land tenure and cultural values due to in-migration	7	10	9	11	6	7
Expansion & intensification of tourism and recreation	8	7	12	10	12	4
Logging or other extractive uses of plants	9	9	5	6	8	6
Construction & maintenance of roadways	10	5	7	5	15	11
Conversion of wildlands to agriculture	11	12	10	12	10	12
Poorly managed fuel loads and fire regimes	12	8	8	7	7	8
Demographic shifts affecting rural & urban ethnic mixes	13	14	13	14	14	13
Influences of mass media and advertising	14	11	15	13	13	15
Influence of global free trade on local communities	15	15	14	15	11	14

Fifty-nine of the 71 respondents were grouped into five categories, based on professional affiliations (12 did not self-identify). Ranks range from most important (1) to least important (15). Additional threats identified by the respondents, and their perceived importance, are discussed in the text.

related to the extraction of water and energy resources to fuel the economic expansion of outlying urban areas. Understanding and explaining this apparent contradiction is not simple. Returning to our three types of threats (magnitude, geographic extent, and immediacy) offers one way to interpret this contradiction.

For the most part, immediate, local threats rank higher than those of broader extent, even among people who offered broad and expansive replies to open-ended questions about the most important issues threatening biocultural values. Dams, overgrazing, invasive plants, and mining rank high, whereas demographic trends, mass media, and economic effects rank low. Threats that can be easily located, examined, and targeted occupy the top spots; those that are harder to describe, quantify, and combat are found at the bottom of the list. These patterns tell us something about how we perceive threats, and clarify our relative concerns. When faced with a task of problem identification, we tend to highlight underlying causes and broad trends. When we turn to prioritizing threats and formulating responses, however, we tend to focus on specifics and discrete problems. Recognizing this dichotomy in the perception of threats, especially among the informed and experienced group of Colorado Plateau environmental professionals, is important, and we shall return to it when we move to a discussion of response options that may help to protect the unique character of the plateau.

Table 2 also reveals a striking similarity in the rankings put forth by different subsets of respondents. Broad agreement is shown regarding the importance of the top five threats, as well as the three lowest ranking threats. Reservation-based respondents were in general agreement with those who work for government agencies, although the devaluation of traditional values ranked much higher for the respondents who work on indigenous lands. This similarity may reveal the pervasive influence of the federal government (through the Bureau of Indian Affairs and other agencies) in Indian country. It is not surprising that environmental activists paid more attention to the effects of tourism and recreation, whereas federal and state agency professionals ranked invasive species—the subjects of recent funding increases meant to control and eradicate such "exotics"—higher than did other groups. Overall, however, the most striking result of the threat rankings is the general agreement among various subsets within the community of environmental professionals regarding the threats generated by increasing human populations, urban sprawl, and water impoundment, diversion, and pumping.

> Life on the high arid plateau became viable when the human beings were able to imagine themselves as sisters and brothers to the badger, antelope, clay, yucca, and sun. Not until they could find a viable relationship to the terrain, the landscape they found themselves in, could they emerge. Only at the moment the requisite balance between human and other was realized could the Pueblo people become a culture.
>
> —Leslie Marmon Silko (1987)

Prioritizing Conservation Efforts

How can we combine our deeper appreciation of the different types of threats with our knowledge of the Colorado Plateau's biocultural diversity, in a manner that will help guide efforts to safeguard the uniqueness of this ecoregion? Too much information about the variety and importance of biological and cultural resources and the processes that sustain them can have the opposite effect; that is, too much detail can obscure, rather than clarify, our perspective on conservation, drowning the most important calls for action in a tidal wave of specific cases of threat and decline. A more panoramic perspective is needed, one that can help us focus on the most pressing issues, identify the critical elements at risk, and locate the specific trends that threaten them. By synthesizing a detailed understanding of "uniqueness" with a clearer perception of the nature of threats, we can highlight the times and places where high value and great threat overlap: these are the priority issues for conservation (Sisk et al. 1994).

The remainder of this publication provides the detailed information and insight needed to assess priorities and evaluate possible responses. But turning once more to the insights of environmental professionals from across the Colorado Plateau, we offer a direction of inquiry, and a means of combining perspectives involving resources and threats for an integrated, synoptic assessment. Figure 3 illustrates a method for plotting the overall importance of biocultural values (along the horizontal axis) against the perceived threats. The graphing space is divided into four quadrants for the purpose of generalizing the results. In the upper right quadrant are the high-priority conservation issues, where highly valued resources face the most critical threats. Based on our practitioner survey, water resources and associated biotic communities are in this highest-priority quadrant of the graph, because rapid urban growth and water development impact them directly. In the lower left quadrant are issues where less-critical resources face less-pervasive or impending threats.

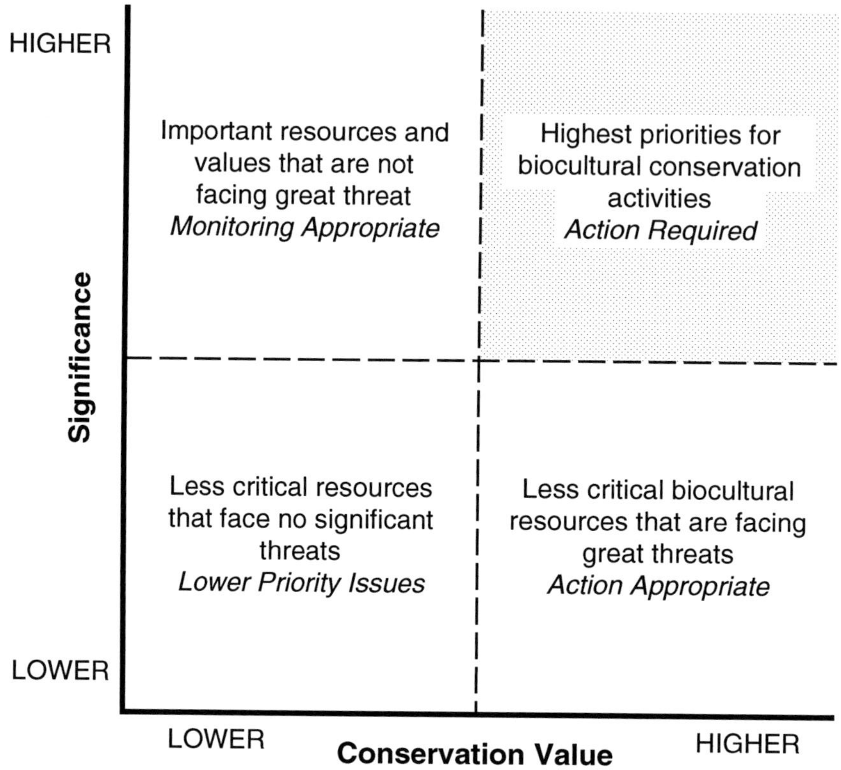

Figure 3. Graphical evaluation of high-priority issues for safeguarding the uniqueness of the Colorado Plateau. Conditions where a high conservation value overlaps with a significant threat constitute top priorities for conservation actions. Lower values facing great threats probably represent the next tier of actions, and key resources not currently at risk should be targeted for ongoing, efficient monitoring efforts so emerging threats can be identified early.

Our survey results suggest that the conservation of many rare terrestrial species might fall into this quadrant, if the trends endangering them were localized or modest in magnitude. This is not to suggest that these issues are not worthy of attention; rather, the opportunities for addressing these problems may be many, or there is sufficient time to address them locally. The rest of the figure identifies issues of intermediate importance, distinguishing between those that concern critical resources facing only modest threats (for example, montane and alpine ecosystems), and resources of lesser importance that are at greater risk (for example, some isolated sagebrush shrublands). This concept for assessing priorities is admittedly simplistic; it cannot begin to encompass the complexity of working with these issues on the ground, in the "real world." However, it does provide a way of distinguishing issues and prioritizing efforts. One insight that emerges from this effort is the recognition that not all threats need be attacked with equal intensity. Change is an integral part of all biocultural systems; yet some types of change, such as the suppression of normal fire regimes in fire-adapted forests, can be devastating (Allen et al. 1998). If detrimental change cannot be stopped, there are clearly cases where it must be redirected.

This framework, grounded in the responses of 80 environmental professionals from across the Colorado Plateau (Table 3),

Table 3. Actions to diminish existing threats to important biocultural resources on the Colorado Plateau, ranked according to their importance in reversing undesirable trends.

Threat	All Respond. (71)	Federal Agencies (19)	State Agencies (7)	Universities (21)	Reservations (2)	Environ. Orgs. (10)
Greater investment in ecological restoration	1	2	1	2	4	4
Greater investment in environmental and place-based education	2	1	3	3	2	2
Stronger economic incentives to reduce resource consumption	3	3	2	1	5	3
Stronger economic incentives for private land stewardship	4	4	7	4	8	1
Greater investment in cultural preservation	5	6	5	5	3	5
Stronger enforcement of existing laws	6	5	8	8	1	6
Stronger recognition of tribal sovereignty	7	10	6	6	6	7
Greater federal control of resources	8	7	11	9	10	10
More capacity-building on and off reservations	9	8	4	7	9	7
Greater local control of resources	10	8	9	10	7	9
Greater state control of resources	11	11	10	11	11	11

Fifty-nine of the 71 responding environmental professionals were grouped into five categories, based on professional affiliations (12 did not self-identify). Ranks range from most important (1) to least important (15). Ties are given the same rank, and the next lowest rank is skipped; all ranks can be compared among categories of respondents.

provides some initial guidance for prioritizing conservation actions. More detailed treatments of key issues follow in subsequent chapters, but we return to this framework—the identification of key biocultural resources and the examination of threats facing them—as we attempt to address, honor, and share the insights gained from a synthesis of scientific research, cultural traditions, and field experience concerning the uniqueness of the Colorado Plateau.

Final Words: The Limits of Perception and the Need for Humility

Humans have seldom been effective at predicting the future. The attempt to anticipate threats is, in fact, an effort to identify negative trends and project them into the future. But what we see as threats today may turn out to be "opportunities" for positive change in the future. Change has often been viewed as a threat, in and of itself, and this perception can hinder creative conservation efforts, locking us into an impossible mission of arresting environmental and cultural change. Historically, however, we have more often been guilty of the opposite error. The embracing of what was initially perceived as positive change—for example, the development of uranium mining on Indian lands, the damming of the Colorado River, and the construction of huge, coal-fired power plants in pristine landscapes—has brought negative consequences for local people and ecosystems. Economic expansion has typically sacrificed local values for regional or national economic growth.

Although we cannot be sure that our analysis and examination of important resource threats is "ahead of its time," we do know that failure to consider existing and emerging threats to biocultural diversity will lead to the erosion of this diversity because, otherwise, the future will be driven by the expanding population and economic activity of the greater Southwest, a region that looks increasingly to the Colorado Plateau for power and water. We cannot claim to know the future before it happens, but we can listen to the lessons of the past and use this knowledge as our context for looking forward in time. Without jumping to conclusions we must make judgments, and understanding the diverse array of negative trends that we call "threats" is essential to making informed decisions about the resources and values we hold dear.

Geographic Heterogeneity: Climate, Stratigraphy, and Soils of the Colorado Plateau

Lawrence E. Stevens and Gary Paul Nabhan

The Colorado Plateau is among the largest arid drainage basins on the North American continent; the Colorado River watershed core of the plateau covers 939,000 sq km between the Rocky Mountains to the east and the Uintahs to the west. Its elevational amplitude is considerable, with localities on the plateau situated as low as 1200 m and as high as 4600 m. This extremely rugged, geologically heterogeneous region has been shaped and fragmented by many factors, including the downcutting of the Colorado River, which descends 3600 m from its headwaters at Mount Richthofen on the Continental Divide in Colorado, to its mouth at the Sea of Cortes 1850 km downstream (Beus and Morales 1990). The Colorado Plateau's large elevational amplitude, its pedogenic heterogeneity, and the high degree of fragmentation of its landforms have left a great variety of habitats for its plants, animals, and people. To grasp how geology has shaped biodiversity and cultural diversity on the Colorado Plateau, we highlight factors that have apparently helped to isolate populations in space and time, contributing to high rates of biological endemism, and ultimately to considerable diversification.

Over the last ten millennia, it appears that humans have visited this entire range of altitudes, although they primarily resided in the 45 percent of the plateau found below 1850 m. Indigenous and immigrant residents alike have regarded as sacred several of the mountain ranges on the plateau that rise above 2500 m, including Navajo Mountain, the San Francisco Peaks, the Chuska Mountains, Mount Taylor, and the White Mountains.

History and Status of the Region's Geological Wonders

You shall not look upon its like again, for the world elsewhere does not show it. The gray-green surface of the earth has been broken through here and the rose and gold of the underlying strata are revealed ... the air itself responds with veilings of lilac and purple.

—John Van Dyke (1920)

The Colorado Plateau is remarkable for having retained a vast stack of igneous, metamorphic, and sedimentary strata in their original depositional order, as they were laid down during the second half of the earth's history (Baars 1983). Relative to the rest of the Rocky Mountain Geologic Province, the Colorado Plateau has remained rather stable from pre-Cambrian times to the present. During the past 2 billion years, every major habitat type on earth has existed in the area where the Colorado Plateau is now located, and many of these paleo-environments have left records of their history recorded in the region's strata. As such, the Colorado Plateau has provided the evolutionary theater for the development of unique forms of flora and fauna, with adaptations to the many spatially and temporally diverse settings found in this region (Carothers and Brown 1991). Present-day species generally appear to be extremely well adapted to the physical environments they inhabit on the plateau. Nevertheless, recent changes in flood discharge regimes, water quality, mean summer daytime temperatures, and competition from exotic species have created relatively novel environmental stresses, challenging the adaptive capacities of many native species.

To profile the amazing variety of rock types and derivative soils present on the Colorado Plateau, we begin with the oldest stratum on the continent west of the Rocky Mountains: the Vishnu Schist, a 1.8 billion-year-old crystalline "basement" bedrock. The strata that overlay the Vishnu Schist provide a record of all subsequent geologic periods except the Orduvician and Silurian. Pre-Cambrian and Paleozoic strata are exposed in the Grand Canyon, whereas Mesozoic and Cenozoic strata are exposed farther north along the "Grand Staircase" (Babcock 1990).

The oldest strata exposed at the bottom of the Grand Canyon, Vishnu Schist and Zoroaster Granite, remind us that thousands of meters of sedimentary and igneous deposits accumulated on the floor of a pre-Cambrian sea prior to 1.8 billion years ago, and these underlie much of the Colorado Plateau. A mountain range came and went, eroded off flat by 1.2 billion years ago. Shallow marine sediments accumulated on the low, rolling peneplain of schist and granite, and lithified into limestone that contains the plateau's oldest fossils—algal stromolite reefs. Few prehistoric or historic inhabitants of the Colorado Plateau ventured into the stretch of the Grand Canyon where these pre-Cambrian strata are exposed, but now 2500 river runners each year marvel over these antiquities as they raft through the Colorado River corridor of the canyon.

> Unroll that topo map again and scrutinize the tablelands dipping back from the canyon toward the San Francisco Peaks. Model makers gather clues to the underground terrain—secrets back from the lab, water levels observed in a handful of wells, data on the Plateau's underpinnings and magnetic fields—and plug them into equations, then fiddle with water flowing in, flowing out.
>
> —Ann Walka (2002)

During the Paleozoic era, a shallow marine shelf stood where sediments forming the Colorado Plateau are located today, dominating the terrain from 550 to 250 million years ago. Cambrian sediments began to develop along and beneath this Cordilleran sea. Both Tapeats Sandstone and Bright Angel Shale beds were left behind by these seas. Additional sediments, which developed into sandstone, limestone, and shale, were deposited during the Mississippian and Pennsylvanian periods. In addition, 4300 m of salt deposits were left behind on the northern Colorado Plateau. The salt and gypsum deposits associated with the Pennsylvanian Hermosa group in the Paradox Basin are exposed at only a few places near the confluence of the Colorado and Green Rivers, but contribute significantly to the high salinity of the river's waters. These deposits have been actively sought out by mineral salt gatherers on the plateau for centuries, if not millennia. In addition, a number of unique plants (edaphic endemics) have adapted to the bizarre geochemical conditions found in Paradox Basin sediments (Baars 1989).

Permian strata record a change from marine deposition of limestones in the west to terrestrial, wind-deposited beach sediments in the east. These sediments developed into Cedar Mesa Sandstone and the mixed strata of the Cutler Formation. They were followed by a variety of late Pennsylvanian and Permian deposits that were more localized in their geographic extent: Hermit Shales, Organ Rock Shales, Coconino Sandstones, Toroweap Limestone, White Rim Limestone, Kaibab Limestone, and possibly, Moenkopi red mudstones (Beus and Morales 1990). Areas where these sediments surface were prehistorically inhabited by small settlements of Anasazi, Fremont, and Salado culture-bearers, who foraged and later farmed on the dunes, mesa edges, and floodplains derived from these sedimentary rocks.

Mesozoic-aged rocks are scarcely visible on the southern half of the Colorado Plateau, but are well exposed in its Canyonlands subregion. There, Mesozoic strata reach up to

1500 m in thickness and can be found in the following (ascending) order: the silty sandstones of the Moenkopi Formation; the lacustrine Chinle Formation; the Glen Canyon Group, including the wind-deposited Wingate and Navajo Sandstones, among other formations; the San Rafael Group, including Page Sandstone and other sedimentary deposits; the mixed Morrison Formation, rich in dinosaur fossils; the Cedar Mountain and Burro Canyon Formations from the Cretaceous; the Mancos Shale, home to more endemic plants than any other stratum on the plateau; and the Mesaverde Group, including shales, sandstones, and coals (Beus and Morales 1990). The alkalinity or acidity, porosity, fertility, and erodability of these sedimentary formations vary greatly, and affect vegetation composition and density in dramatic ways. For example, the presence of shales on an arid, south-facing slope may allow the persistence of desert scrub plants 1200 m higher than they are found on adjacent north-facing slopes of soils derived from sandstones (Nabhan and Wilson 1995). In short, soil and rock type, combined with slope, aspect, and elevation, create complex mosaics of vegetation wherever a mix of Mesozoic sediments are exposed on the Colorado Plateau. Because Colorado Plateau soils may take 5,000–10,000 years to develop in this arid climate, many areas of the plateau remain exposed down to their bedrock parent material. For this reason, the Colorado Plateau is reputed to have the largest expanse of relatively barren slickrock badlands of any region on the continent. However, even these badlands are not truly barren and lifeless. Many of the slickrock sandstone benches on the plateau harbor endolithic algae, which provide nutrients to the other sparsely distributed life forms in the area (Nabhan and Wilson 1995).

The Cenozoic, Tertiary, and Quaternary periods on the Colorado Plateau included periods of intense volcanism, which left behind vast lava flows, cinder cones, basalt ridges, tufaceous hills, and plains covered by cinder mulches. As recently as 800 years ago, volcanism in the Sunset Crater area left behind fertile, potassium-rich mulch attractive to Sinagua farmers, who settled in the area until field soil fertility was apparently impoverished from several decades of intensive cultivation. Volcanic features also offered caves for habitation, blanks for grinding stones and other tools, and cobbles for terraces and check-dams.

A variety of other geological processes have led to the fragmentation of the plateau: folding, uplifting, faulting, downcutting, and the blocking of paleo-streambeds by volcanic flows, lava dams, and silt plugs (Hamblin 1990). The dizzying diversity of mesas, buttes, cinder cones, calderas, canyons, and cliff-faces on the plateau provided resident cultures with a multitude of building materials, sites for excavating caves, cliff dwellings, storage niches, and strategic positions for defense. Overall, the geographic heterogeneity of the region allowed language groups to become fragmented into subgroups of people who developed distinctive dialects, and in a few cases mutually unintelligible languages of their own. Different cultures also began to specialize in the use of certain sets of geomorphic habitats, with their attendant flows of water, seeds, worms, fish, or fowl.

Trends and Scenarios

Public appreciation for the geological splendor of the Colorado Plateau has led to the formal protection of numerous canyons, natural bridges, arches, and geomorphological anomalies such as "hoodoos" (columns of rock shaped by erosional processes into distinctive figurines) over the last century. More than 10 million tourists per year visit these geological wonders, and many of them remain sacred places to indigenous inhabitants of the region: Rainbow Bridge, Paalatsomo or Sunset Crater, Munkuntuweap in Zion National Park, Canyon de Chelly, the Grand Falls of the Little Colorado River, or the narrow, travertine-laden falls in Havasupai Canyon (Milne 1995). However, increasing tourism and recreational use around canyons, arches, natural bridges, and spires can have devastating effects. The most flagrant known case of profaning a geological icon occurred within the last decade, when nature photography tour leader Michael Fatali scorched the bottom of Delicate Arch in Arches National Monument by setting chemical fires beneath it to "artificially light the landscape" as a special effect (High Country News 2000). He was charged with seven counts of misdemeanor for his degradation of this landform.

> The most sacred of places is made powerful by the history, stories, songs and prayers it contains. As we see this place, it is an experience of awe and gratitude. It is as if the Holy People are physically comforting us, encouraging us, smiling at us, strengthening us. That Diné Tah seems an empty, barren place suits us—we are among the most fortunate people in the world because of it.
>
> —Luci Tapahonso (1995)

On a larger scale, the world's largest stone arch has been degraded by the flooding of Glen Canyon, which created a National Recreation Area around Lake Powell, and which brings as many as a thousand boaters per day up to Rainbow Bridge. For two decades, there has been concern that as Lake Powell reached its highest level, increased water erosion could wear at the base of the bridge and threaten it. In addition, Lamarrr Badoni and other Navajo medicinemen have attempted to prohibit public access to the bridge and reduce Lake Powell's volume to half of its storage capacity out of fear that the flooding of Bridge Canyon would inundate a holy water spring—a prayer spot—and weaken Rainbow Bridge, a landform sacred to the Navajo, Hopi, San Juan Southern Paiute, Kaibab Southern Paiute, and White Mesa Ute. Despite their efforts, they lost their case in a 1980 ruling of the 10th U.S. Circuit Court of Appeals, on a technicality: Rainbow Bridge was designated as a National Monument for recreational visitation 10 years prior to Navajo Reservation expansion into the area, and before the federal government's nation-to-nation recognition of the Native American cultures' tribal councils (Smith and Manning 1997). More recently, the Mountain States Legal Foundation has challenged the legality of National Park Service signage that suggests that non-Indian visitors to Rainbow Bridge should refrain from stepping under the bridge out of deference to the religious concerns of the Diné Medicinemen's Association. Should Mountain States Legal Foundation continue to assert that protection of sacred geological features is unconstitutional, overvisitation and degradation of sites will inevitably be accelerated.

> The question, then, is why should a butte that is sacred to Native Americans be treated differently than a mission or a church managed by the Park Service? Because the butte is a natural feature and Native American religions are harder to understand than Western ones?
>
> —Chris Smith & Elizabeth Manning (1997)

Prospects and Opportunities

Fortunately, there are a number of coalitions involving scientists, resource managers, and activists involved in safeguarding the diverse geological features of the Colorado Plateau, from the world's largest stone arch to microscopic crusts. With regard to Rainbow Bridge, nonprofit groups such as the Sacred Lands Film Project, Living Rivers, and its Glen Canyon Action Network have all supported and assisted the Diné Medicinemen's Association with their efforts to minimize degradation and desecration of Rainbow Bridge. The Sacred Lands Film Project Web site also informs activists of similar issues affecting geological features of multicultural significance such as Woodruff Butte, Black Mesa, Zuni Salt Lake, and Petroglyph National Monument.

Perhaps the most unique conservation effort to emerge out of the Colorado Plateau is the educational effort to protect and restore the ancient soil crusts that represent 70 percent of the ground cover at lower elevations in this ecoregion. These soil crusts are both geological and biological, gaining their microstructure from symbioses among cyanobacteria, lichens, and mosses glomming onto sand and silt particles. Also known as cryptogamic, cryptobiotic, or microphytic crusts, these microscopic features stabilize otherwise erodable soils, increase water filtration and fertility, and provide safe microsites that are important for the seed germination of annual desert wildflowers and shrubs (Harper and Marble 1988; Johansen 1993; Belnap and Gardner 1993). When it became clear in the 1960s that these soil crusts can be easily damaged by off-road vehicles and livestock herds, and may take from 50 to 250 years to recover, the National Park Service staff at Arches National Monument began public education to reduce off-trail as well as off-road impacts. Over the last four decades, environmental educators in parks, monuments, and schools across the entire Colorado Plateau ecoregion have informed millions of visitors and local residents that these crusty microgeological features are what "hold the place in place," and that the plateau is endowed with some of the most well developed crusts known anywhere in the world. In fact, this may be one of the first and certainly the most successful "soil and microbial rights to exist" campaign on the planet. The move indicates vital, future-conscious ways of conceptualizing and interacting responsibly with the landscapes on which we live.

Hydrological Diversity: Water's Role in Shaping Natural and Cultural Diversity on the Colorado Plateau

Lawrence E. Stevens and Gary Paul Nabhan

Water, although naturally scarce on the Colorado Plateau, is perhaps the most powerful and dramatic shaper of this ecoregion's landscapes and cultures. Rain, or more properly the lack of and hope for it, is the focus of many indigenous ceremonies on the plateau, from the Hopi Bean Dance to the Corn Dances of the New Mexico Pueblos. As in the Odyssey, water is the medium on which we humans sail between or wreck on the Charybdis and Scylla of resource exploitation or stewardship, toward or away from home.

Because the Colorado Plateau is essentially an arid and semi-arid plateau where precipitation in lowland areas can be as little as 10 cm per year, water is often the limiting factor for the growth of plant life, animal life, and human enterprises. The highest densities of human and nonhuman populations on the plateau are clustered around its scarce perennial sources of fresh water—rivers flowing in from adjacent uplands, small creeks and streams, and artesian springs. For example, there is sometimes 500 times the number of species at springs than in the surrounding arid landscape, and spring-fed wetlands are often 1000 times more productive than the desert vegetation just upslope from them. It is not surprising, then, that the needs of wildlife and human communities are often politicized and pitted against one another, as growing towns and cities have usurped the hydrological resources that are essential for maintaining biodiversity. In the end, however, resolving water allocation disputes in ways that meet the needs of various life forms and cultural communities can ultimately remind us of our interdependence.

> Water is like glue—it has the capacity to bring us all together.
>
> —Vernon Masayesva (2002)

To describe the vital role of water in shaping the natural and cultural uniqueness of the Colorado Plateau, we first highlight the region's dramatic variation in precipitation, evaporation, surface hydrology, and groundwater hydrology. As part of this discussion, we report some of the diverse biotic and cultural adaptations to the uneven temporal and spatial distribution of water resources that characterize the Colorado Plateau. In addition, we summarize the current status of the plateau's water supply, as well as some of the scenarios that may occur if current declines in water quantity and quality continue to threaten the ecoregion's natural and cultural diversity. We also note how cross-cultural water politics have changed during the last half century; in many cases, the trends have been toward reaffirming the rights of the Colorado Plateau's native nations. As we shall see, almost without exception these indigenous communities have historically utilized limited water resources in ways that helped conserve aquatic biodiversity.

Status and Trends of Precipitation and Evaporation

Because the prevailing winds reaching the Colorado Plateau are westerly, moving inland from the Pacific Coast to the continental interior, orographic precipitation is often greater on west-facing flanks of mountains such as the Wasatch Front, the White Mountains, or the San Francisco Peaks. Consequently, much of the Colorado Plateau lies in a partial rainshadow created by the Sierra Nevada, the Wasatch Range, and other mountain ranges farther west. As a result of steep elevational gradients, orographic lifting, and other geographic factors, precipitation on the Colorado Plateau ranges from less than 100 mm at sites fully within rainshadows and below 1400 m, to nearly 1000 mm per year in mountainous regions above 2500 m (Windell 1986). In a single watershed such as that of the Paria River, there may be as much as 420 mm of precipitation annually (including 3 m of snow) in the high-elevation headwaters, but only 50 mm of mean annual rainfall where the Paria flows into the Colorado River (Hereford 1984). In addition to this spatial variability in rainfall, the seasonal influence of six kinds of storm systems creates considerable temporal variation in moisture patterns (Hansen et al. 1977).

Because the Colorado Plateau's overall precipitation pattern is bimodal—with winter storms coming from westerly frontal systems, and torrential summer monsoons developing in late summer—different guilds of annual wildflowers and solitary bees have evolved to take advantage of seasonal pulses in moisture availability. The early spring pulse is dominated by plant species derived from the Mohave and Great Basin Deserts, whereas the later summer-autumn pulse has more overlap with the Sonoran, Chihuahuan, and Great Plains floras. Arid regions with bimodal precipitation patterns tend to have greater species richness of ephemeral herbs, grasses, and insects than adjacent regions dominated by either winter or summer rains (Felger and Nabhan 1976).

Catastrophic storms also erratically shape the landscape, its watercourses, and its perennial vegetation. One such storm deposited 89 mm (3.5 inches) of rain in Mesa Verde National Park in 45 minutes on August 3, 1924 (Hansen et al. 1977), generating considerable floods on the Ute and Navajo lands below the park. Hailstorms are also common during the summer monsoons, with hailstones up to 4.5 cm in diameter reported to have damaged fields, orchards, and native perennial vegetation (Rykaczewski 1981). Snow can occur from October through May at higher elevations, limiting the northern and upland distribution of many arid-adapted plants. Total snow accumulations exceed 5 m/yr at higher elevations, and snowmelt in springtime provides enough runoff for rivers to surge and for some seasonal flow rates of springs to increase dramatically. The prehistoric and historic farmers of the region timed their plantings to take advantage of these surges in the flow of water; elsewhere, they planted their crops in areas where residual moisture was left over from snowy winters, allowing the plants to survive until the summer monsoons began.

Particularly at lower desert elevations, evapotranspiration rates exceed precipitation in the early summer before the monsoons begin. The arid portions of the plateau suffer pan-evaporation rates in excess of 2.3 m/yr, and many areas have aggravated evapotranspiration due to the "clothesline effect." This process can be driven by hot, dry winds moving across kilometers of black volcanic cinders, pale dunes, or slickrock before encountering a field, wetland meadow, riparian gallery forest, or park-like pasture, causing moisture to be wicked out of the soil and vegetation rapidly. Under such conditions, cottonwoods, tamarisks, and mesquites on the plateau can transpire the weight of their foliage every hour at temperatures of 30° C (Davenport 1978). However, cottonwoods and willows located around the edges of fields may help to reduce moisture stress in adjacent croplands by protecting crops from desiccating winds, by increasing humidity and shade, and by reducing ambient temperatures. Noticing these benefits, certain Native and Hispanic farmers maintained or even transplanted these riparian trees to serve as hedgerows and windbreaks adjacent to their crops (Whiting 1942; Sheridan and Nabhan 1979).

Trends in the precipitation balance on the

Colorado Plateau suggest that residents have begun to face drier, hotter years with greater frequency, which will likely trigger more crown-burning wildfires in forests and woodlands where fires have been largely suppressed for a century or so. When the last period of prolonged summer droughts on the Colorado Plateau occurred from 1130 to 1180 A.D., it dramatically reshaped Anasazi and Sinagua settlement patterns and population densities (Euler et al. 1979).

In more recent times, a severe drought in the 1950s depleted rangeland cover and decimated livestock herds across the Colorado Plateau. Nevertheless, tree ring data taken from Malpais National Monument on the southeastern edge of the plateau suggest that the past two centuries have been the wettest period of the past 1500 years. These data also suggest that the period from 1980 to 2000 has been among the wettest of all, with rainfall 23 percent above the long-term regional average (Grissino-Mayer et al. 1997).

In other words, the rapid demographic growth of the last two decades on the plateau has occurred during an atypical period during which urban and suburban development and agricultural irrigation networks of the region had access to rivers, springs, and reservoirs swollen by unusually wet weather. It is not very likely that this unusually wet period will extend into the new millennium, if only because the long-term global warming trend that became evident in the 1960s is likely to intensify, further drying soils and reservoirs, increasing evapotranspiration, depleting vegetation, and increasing surface albido levels. These factors all have some indirect influence on decreasing the probability of local thunderstorm build-up, although the net effect may be more unpredictable rainfall rather than less rainfall. Because global surface temperatures reached a record-setting high level in 1998, and 1999 was the sixth warmest year on record, we should begin to see the time-lag effects of these changes on bird and mammal life as well. Just east of the Colorado Plateau at Gothic, Colorado, ecologists have already documented a 14-day earlier arrival time for migratory robins than in 1981, and a 38-day earlier emergence time of yellow-bellied marmots than in 1977 (Inouye et al. 2000).

Status and Trends of Surface Waters

Despite its many political and cultural subdivisions, the Colorado Plateau functions as a single isolated hydrological drainage entity. As a result of this isolation, more than 68 percent of the Colorado River's fish are endemic, found in no other river basin. The Colorado Plateau is fed by snow melt and springs, and the Colorado River receives most of its water from winter precipitation in peripheral mountains. Iorns et al. (1965) estimated that approximately 92.7 million acre-feet (maf) of water entered the upper Colorado river basin as precipitation annually. Of this amount, 2.3 maf (or 2.4%) were consumed or diverted, 12.7 maf (13.7%) left via surface outflow, and 77.7 maf (83.8%) were lost to evaporation and transpiration. Annual consumptive use had increased to 6.6 maf by 1980 (U.S. Department of the Interior 1987), with more than 55 percent of the consumptive use going to transbasin diversions, 30 percent to agriculture, 11.5 percent to evaporation in reservoirs and channels, 2.7 percent to municipal and industrial use, and none for fish, wildlife, or recreational uses on the Colorado Plateau (Iorns et al. 1965).

Price and Arnow (1974) estimated that phreatophytes—deep-rooted trees such as cottonwoods which "pump" groundwater to assure their growth—in the Colorado River basin consumed 2.5 maf of ground and surface water annually, which is the equivalent of municipal and industrial consumption.

Despite the apparent abundance of water on the Colorado Plateau, groundwater quality is often less than acceptable. The Paradox Basin in eastern Utah and western Colorado is underlain by strata containing sodium chloride and potassium, and as much as 70 percent of the recoverable groundwater in the upper Colorado River Basin is saline and either detrimental to, or of limited value for agricultural, municipal, or industrial uses. For example, potable groundwater was found down to 1980 m depth in the White

River drainage, but salinity was high below that depth, with 2000–80,000 ppm TDS (total dissolved solids) and 10 times the acceptable fluorine concentration (Masse and Adams 1983). Calcium and magnesium bicarbonate ions in groundwater derived from Quaternary alluvium had 700–25,000 ppm TDS, and groundwater from the Tertiary Green River formation contained up to 12,000 ppm TDS.

Many of the settlements of the prehistoric and historic farming cultures of the Colorado Plateau were clustered along the perennial and intermittent tributaries of the Colorado River to take advantage of more reliable access to fresh water. The Navajo, Hopi, Southern Paiute, Havasupai, Yavapai, Western Apache, Laguna, and Acoma Tribes are among the indigenous cultures who either diverted or terraced surface flows with check-dams. Because of frequent natural flooding—with some of these natural occurrences greater than the 20–33 cubic meters per second (cms) "managed floods" of the 1990s—the banks of the Colorado itself and some of its larger tributaries were less suitable for farming. In contrast, the greatest flood on record for the Colorado River occurred in 1884, perhaps resulting from the climatic shifts generated by the Krakatoa volcanic eruption in the South Pacific, approaching 8500 cms.

> The unimpeded Colorado River eroded [vast quantities of] sediments, efficiently moving them on their journey to the sea. Dam construction corrupted this cleansing system, creating sediment traps in the reservoirs. Lake Powell traps thousands of tons of sediment every year from the Upper Colorado River basin, which according to the Department of the Interior, harbors some of the worst water quality conditions in the nation due to the sediments that are being farmed and eroded.
>
> —David Wegner (1998)

The history of the Intermountain West since that time has been shaped more by the development of a technologically based hydraulic civilization than by any other factor; dams have directed the region's destiny (Worster 1985). Since the 1880s, more than a hundred moderate to large-sized dams have regulated the flow of the Colorado River, from near its headwaters in the Colorado Rockies to its mouth at the Gulf of California. Humans now consume each gallon of the river's water at least three times before it leaves the Colorado Plateau (Fradkin 1984; Reisner 1993). Water has come to be equated with power on the Colorado Plateau, both literally and figuratively: river flows have been used to generate 1823 megawatts per hour of hydroelectric energy from the upper Colorado River basin, which provides about 3 percent of the total power used on the plateau.

At the same time, the building of dams and reservoirs has meant that several indigenous cultures have lost access to many of their farming and fishing spots along the tributaries of the Colorado River. The overallocations of surface flows in the Colorado River basin this century have led to considerable cultural conflict and a plethora of lawsuits, pitting rural Indian and Hispanic communities against many mining companies and municipalities (Pontius 1997). Although Native Americans in the Colorado River basin have already been awarded nearly 1 million acre-feet of water per year in settlements of tribal water rights since 1978 (Pontius 1997), more than 2.5 maf remain contested in the same basin.

> Your sovereignty flows from your water. Go out there and relocate the places your tribal elders named along water courses, then find what basketry, ceremonial, and medicinal plants grow around those streams and springs. They can help you re-establish your sovereignty.
>
> —Chris Coder (2002)

Plausible Scenarios and Future Prospects

Depending upon how these cultural conflicts are settled, the control of water rights and of power generation may dramatically shift. Pending Indian water rights cases affecting nearly 2 maf per year may result in further flow reduction if upper basin tribes are granted rights to control more water for agriculture or for fish and wildlife-based recreation (Pontius 1997).

The outcome of these settlements is sure to affect not only the health and prosperity of cultural communities on the plateau, but their interconnections with biodiversity as well. Recently the Working Group on the Endangered Species Act and Indian Water Rights (2001) completed a case study of the effects of Navajo Dam on the San Juan River and the Flaming Gorge Dam on the Green River. Navajo Dam was built for the potential benefit of the Navajo, Ute Mountain Ute, Southern Ute, Jicarilla Apache, and Uintah and Ouray Ute Tribes. The Flaming Gorge Dam influences the Uintah and Ouray Reservation as well. However, both dams fundamentally changed the hydrological regimes of the rivers running through these Indian communities, dramatically reducing the population viability of the Colorado River pikeminnow, the razorback sucker, the bonytail chub, and the humpback chub, all of which are now threatened or endangered. As a result, Indian communities have essentially lost access to the native fisheries indefinitely—either for their own use or for attracting sport fishermen from elsewhere. They are also losing access to moving, healing water—more than 700,000 acre-feet per year evaporates from the relatively stagnant surface of Lake Powell (Wegner 1998).

One refreshing counter-trend to emerge over the last decade is the possibility of removal of nearly 500 dams from rivers in the United States, including several in the Intermountain West. The potential value of removing dams has not only engaged Earth First!-style activists, but hydrologists, biologists, and environmental policy experts as well, who are evaluating such proposals at every scale, from silted-in earthen dams in the headwaters of smaller watersheds to the most massive icon in the region, Glen Canyon Dam itself. Robert Glennon of the University of Arizona College of Law thought that the proposal to deactivate Glen Canyon Dam was plausible enough to convene a 2-day conference of lawyers, engineers, scientists, bureaucrats, businessmen, and a former U.S. Secretary of the Interior to fully consider and debate this scenario (Wegner 2000; Carothers and House 2000)—one dismissed by many people as darkly comical but implausible when Edward Abbey first published *The Monkey Wrench Gang* in 1975.

> The public is now learning that we have paid a steadily accumulating price for these [dam] projects in the form of: fish spawning runs destroyed, downstream rivers altered by changes in temperature, unnatural nutrient load and seasonal flows, wedges of sediment piling up behind structures, and delta wetlands degraded by lack of fresh water and saltwater intrusion. Rivers are always on the move and their inhabitants know no boundaries; salmon and shad do not read maps, only streams.
>
> —Bruce Babbitt (1998)

Status and Trends in Groundwater and Freshwater Springs

Although Thomas (1952) considered the Colorado Plateau a single groundwater region, more recent work has divided it into several basins, each with complex, interbedded sequences of strata with multilayered aquifers (Heath 1984). Most of the groundwater recharge in the basin occurs in upland areas, with discharge through fractures and surface springs. An estimated 141.9 cubic km (115 maf) of groundwater is stored in the upper 30 m of the basin, which is about four times more than the storage capacity of all of the man-made surface reservoirs in the basin (Iorns et al. 1965). Of the 114.4 cu km of precipitation entering the plateau, an estimated 4.93 cu km recharges these multilayered groundwater reserves (Price and Arnow 1974). When precipitation enters the ground on the Colorado Plateau, it may end up traveling down thousands of meters vertically, and up to 500 km laterally over periods of up to 10,000 years. As a result, much of the Colorado Plateau's fossil groundwater reserves are situated in deep aquifers that are relatively inaccessible.

> A Havasupai man reminds me that water flowing in springs from the pure center of the earth cycles through us all, touching us—the two-leggeds and the plants and the animal people—like the wind does. To maintain the natural balance, he says, springs must persist, untainted, undistorted, undiminished.
>
> —Ann Walka (2002)

Along the South Rim of the Grand Canyon, for instance, the water table is typically 650 m below the land surface, recharges slowly, and is inextricably tied to the flow of springs in Supai Canyon, Indian Gardens, and elsewhere. Grand Canyon National Park has therefore chosen to satisfy its visitors' thirst by transporting 1.5 cu m per minute of water from Roaring Springs 15 km away in a transcanyon pipeline to reach the National Park Service visitors' centers, staff living quarters, hotels, and concessionaires (Ingraham et al. 2001). The ultimate fate of this water is the Grand Canyon sewage treatment plant, which discharges nearly 500,000 cu m per year into the Clearwell Overflow, draining into an otherwise dry wash along the Bright Angel Fault.

Fractures and seeps flowing out between impermeable strata have created more than 5000 known springs on the Colorado Plateau; Arizona has the second highest density of springs in the United States, averaging 0.015 springs per sq km (Figure 4). According to USGS data, about 22 percent of the springs on the plateau are on Indian land. According to Hopi, Navajo, Jemez, and Zuni elders, there are many more springs on Indian lands than USGS surveyors have enumerated. Relatively high densities of springs and seep-generated hanging gardens occur in the upper Green River, the Price River, the Colorado River near Moab, the Hopi and Navajo lands edging Black Mesa, and along the Mogollon Rim, especially in the White Mountains.

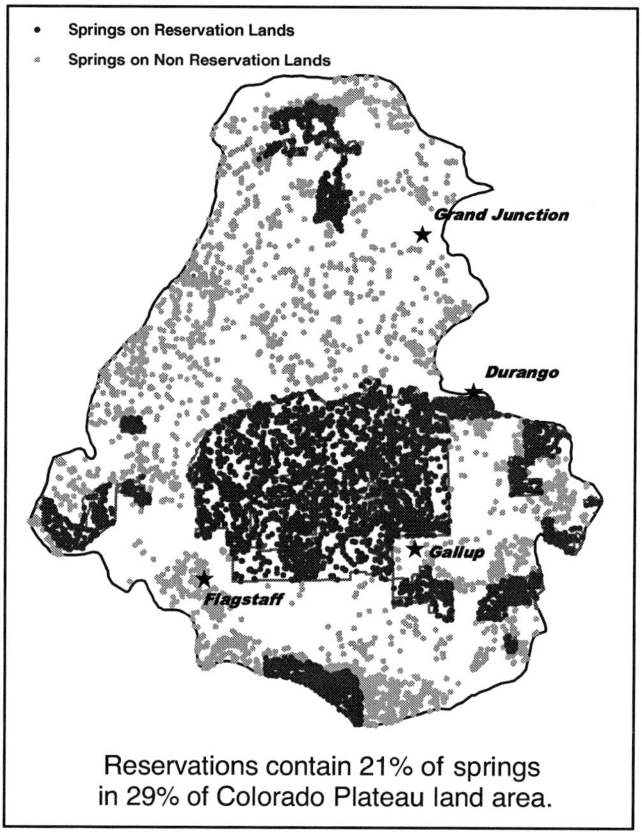

Figure 4. Springs reported since the 1930s on the Colorado Plateau (map by Conservation Technologies Institute, using USGS data).

Unfortunately, an estimated 75 percent of springs in the southern half of the Colorado Plateau have been degraded by human activities, including channeling to water livestock, irrigate pastures, and supply urban residents with water for household use and outdoor recreation. The consumptive water use of invasive species such as tamarisks and Russian olives has also diminished spring flows.

Notwithstanding state and federal regulations that limit pumping by groundwater users to "a safe yield," aquifers under the Colorado Plateau are also being depleted at rates that far exceed recharge. Of recent concern is Peabody Coal Company's pumping of more than 4.9 million cubic meters (1.3 billion gallons) of pristine groundwater each year from Black Mesa to slurry coal to the Mohave Generating Station near Laughlin, Nevada, which produces electricity for Phoenix, Las Vegas, and Los Angeles. Since Peabody's pumping of the Navajo aquifer began, water levels in several wells located on Navajo and Hopi lands have decreased by more than 100 feet. Artesian spring discharge has diminished by more than 50 percent in the majority of USGS-monitored springs in the area (National Resources Defense Council 2001).

> Safe yield is the most dangerous term scientists have come up with. Models projecting what safe yield should be cannot tell you whether a particular spring will dry up due to pumping nearby. We must move to the concept of sustainable yield—factoring in not only human but ecosystem requirements as well, and including what is safe from a cultural perspective.
>
> —Abe Springer (2002)

The decline of these springs potentially affects a number of aquatic and wetland plants and animals. Some of these are used by local tribes in ceremonies or in medicinal, culinary, and material ways. This loss of aquatic habitat around springs potentially affects several Native American religious practices and ceremonial plant uses that are protected by law under the American Indian Religious Freedom Act.

> The Hopi believe springs are like breathing holes. Springs are considered homes of the Water Serpent, Paaloloqangw, and therefore we are not allowed to play around there or even be present near them at noon when he comes out. When the wetlands around springs begin to decline, they are acting like alarm systems, as signs that tell us we're going to be facing some pretty serious problems.
>
> —Vernon Masayesva (2002)

Fortunately, there are now several broad multicultural collaborations working to change policies and to foster cultural equity in the use of groundwater and springs. One such coalition that is now being organized by the Black Mesa Trust, the National Resources Defense Council, the Glen Canyon Institute, the Grand Canyon Trust, and the Sierra Club Environmental Justice Project offers considerable hope for the future. It is exemplary in the way that it synthesizes credible scientific analyses with as much consideration as possible of the long-term value of springs for maintaining wildlife and for indigenous ceremonial obligations.

It is clear that the ongoing trends in Colorado Plateau water management appear to threaten both biological and cultural diversity, as well as the interaction between the two, yet there are more and more hopeful countertrends. The Zuni, White Mountain Apache, Hopi, Santa Clara Pueblo, Jemez Pueblo, Havasupai, Hualapai, and Kaibab Paiute Tribes have all initiated promising programs of riparian habitat protection, wetlands restoration, and aquatic wildlife recovery. In particular, the Zuni Wildlife Department has successfully reintroduced beavers into the Zuni Mountains, helping to revive cultural traditions connected to them (Albert et al. 2000). The formerly frayed threads which once connected water, wildlife, and healthy cultural communities are beginning to be rewoven.

Biodiversity: Plant and Animal Endemism, Biotic Associations, and Unique Habitat Mosaics in Living Landscapes

Lawrence E. Stevens and Gary Paul Nabhan

Traverse the Colorado Plateau from top to bottom, or west to east, and your senses will be bombarded with innumerable signals that this is an ecoregion of dizzying diversity. Climb up from the southernmost reaches of the Grand Canyon river corridor to the North Rim, and you will move from the cascading call of the canyon wren to the clear, questioning notes of Townsend's solitaire in the boreal forest openings of the Kaibab Plateau; your eyes adjust from the pale tans and grays to dark greens; and your nostrils flare with the aromatic lemon-mint oils of Aloysia, then calm to the musky odor of rotting pine logs on the upper edges of wetland meadows. The same is true when moving from the edges of the Mohave and Sonoran Deserts near St. George, Utah, eastward, up into the Colorado Rockies, clear to the alpine tundra, where snowmelt drains into the headwaters of the Colorado River.

This great diversity in biota and heterogeneity of habitats was apparent to C. Hart Merriam and his expedition members soon after they arrived in Flagstaff, Arizona on July 26, 1889. For the next 2 months, these pioneering ecologists completed extensive field surveys in the high forests of the San Francisco Peaks, in the Painted Desert scrublands to the east, and deep into the Grand Canyon. The gradient of biodiversity documented by Merriam and his party in this relatively small geographic area led to his first publications delineating "life zones" on a regional scale (Merriam and Steineger 1890; Brown 1982). Six of the zones defined by Merriam occur in relatively close proximity on the southern Colorado Plateau, from the hot depths of the lower canyons, to the windswept tundra atop the San Francisco Peaks. As most ecologists and amateur naturalists in North America have learned, Merriam used his work on San Francisco Mountain to extrapolate life zones for all of North America, emphasizing that his work on the Colorado Plateau was more significant if placed in the context of the entire continent. Over the following century, the field of ecology grew rapidly, and many of its practitioners used Merriam's maps delineating the major life zones of the continent—maps based on his experience of the "telescoped" diversity gradients found on the Colorado Plateau (Table 4).

The overall elevational amplitude of the Colorado Plateau is from 350 m to nearly 4000 m above sea level, producing within relatively short distances (10–100 km) the entire range of life zones found from Mexico to Canada (Merriam and Steineger 1890). Although our understanding of these ecological life zones has been refined since Merriam first described them, they are still appropriate for teaching students about geographic influences on habitat diversity (Brown 1982; Nabhan and Wilson 1995).

Today, the most widely used system for describing the diversity of biotic communities on the Colorado Plateau is that of Brown and Lowe (Brown 1982). From its highest to very lowest elevations, the Colorado Plateau ecoregion holds examples of nine distinct biotic communities that are defined at levels roughly parallel to Merriam's life zones. However, more detailed classifications of

Table 4. Merriam's life zones (Merriam and Steineger 1890) correlated with Brown and Lowe's (Brown 1982) biotic communities found on the Colorado Plateau.

Elevation	Mean Annual Rainfall	Merriam 's Life Zones	Brown and Lowe's Biotic Communities
1070–sea level	30–7.5 cm	Lower Sonoran	Desert scrub
1980–1070 m	51–25 cm	Upper Sonoran	Coniferous pygmy woodland of pinyon and juniper, semi-arid grassland, interior chaparral
2450–1830 m	66–46 cm	Transition	Petran ponderosa pine forest and associated scrub
2900–2450 m	76–63 cm	Canadian	Mixed conifer forest
3500–2895 m	89–76 cm	Hudsonian	Subalpine conifer forest of spruce and fir
3870–3500 m	101–89 cm	Arctic-Alpine	Alpine tundra

the region's vegetation show that the number of particular biotic (5-digit) associations found on the Colorado Plateau (21) is greater than the number of biotic associations found in the adjacent Sonoran Desert (15) or Rocky Mountain (20) ecoregions (Table 5). This heterogeneity of biotic associations—many of which are closely juxtaposed within the same landscape—is the hallmark of biological diversity for the Colorado Plateau (Harper et al. 1994).

The Colorado Plateau is influenced biogeographically by the biotas of the Rocky Mountain (Petran) biotic province, and by that of the four great American deserts: Chihuahuan, Great Basin, Mohave, and Sonoran. When we attempt to interpret the relative importance of geographic factors contributing to the diversity and uniqueness of the plateau's biotas, two factors and the interactions between them are apparent. The key geographic factor driving diversity and endemism on the plateau is elevation, which controls climate and interacts with a secondary factor—the heterogeneity of soil and rock substrates—in unique and unforeseen ways. A classical example of how these two interacting factors increase habitat heterogeneity can be found in the La Sal Mountains region of southeastern Utah. There, typical desert scrub communities can be found on sands derived from Navajo Sandstone at the base of the La Sals, especially on north-facing slopes and plains. However, on very arid soils derived from Mancos Shale, some of the same plants that dominate desert scrub—saltbush, black brush, and juniper—can be found up to 2500 m on nearby south-facing slopes (Nabhan and Wilson 1995). These high-elevation desert scrub habitats typically have different understory plants such as herbs and grasses, and attract a different assemblage of terrestrial insects and birds. Differences in soil chemistry, moisture-holding capacity, and solar orientation (aspect) have created a 500 m gap between these two desert scrub communities, an elevational amplitude greater than the average vertical breadth of most plant communities found on the same soil and aspect. In such a manner, slope aspect and elevation on the plateau combine to increase the diversity of physical conditions for plant and animal growth and adaptation (Comstock and Ehleringer 1992).

These biotic communities are composed of numerous species that do not "move" together in space and time; that is to say, collectively, they are historically loose associations, and not all of the species found in the region are associated with biotic communities unique to the Colorado Plateau. Nevertheless, it is clear that the Colorado Plateau ranks among the top five ecoregions

Table 5. Terrestrial biotic associations on the Colorado Plateau (following Brown and Lowe, in Brown 1982).

1.111.5	Rocky Mountain Alpine Tundra
121.3	Rocky Mountain Subalpine Forest and Woodland
121.31	Engelmann Spruce – Alpine Fir Series
122.3	Rocky Mountain Montane Coniferous Forest
122.31	Douglas Fir – White Fir Series
122.32	Ponderosa Pine Series
122.33	Gambel Oak Series
122.4	Great Basin Conifer Woodland
122.41	Pinyon – Juniper Series
132.1	Great Basin Montana Scrub
132.11	Oak Scrub Series
132.12	Mountain Mahogany Series
132.13	Maple-Scrub Series
132.14	Bitterbush Series
132.15	Serviceberry Series
132.16	Mixed Deciduous Series
133.3	Interior Chaparral
133.31	Scrub Oak Series
133.32	Manzanita Series
133.33	Ceanothus Series
133.34	Mountain Mahogany Series
133.35	Silk-tassel Series
133.36	Mixed Evergreen Sclerophyll Series
142.2	Great Basin Scrub-Grassland
142.23	Ricegrass Series
143.1	Scrub-Grassland
143.11	Grama Grass-Scrub Series
154.1	Sonoran Desert Scrub
154.12	Palo Verde–Mixed Cacti Series
154.17	Saltbush Series

of North America in terms of its species richness in various taxonomic groups (Ricketts et al. 1999b: Table 3.3). In addition to the documented high species richness for plants, tiger beetles, butterflies, reptiles, and mammals, there are preliminary indications that the plateau also harbors relatively high numbers of ants, grasshoppers, and leafhoppers (Table 6; Nelson 1994; Sigler and Sigler 1994).

Despite its visual wonders, the Colorado Plateau's most legitimate claim to fame may be its high rates of endemism, ranking in the top three ecoregions on the continent for total number of endemics in all taxonomic groups (Ricketts et al. 1999b). For example, it has the greatest number of vascular plant endemics in North America; 290–305 of its plant taxa occur nowhere else on this continent (Kartesz and Farsted 1999; Appendix 2). Many of these are edaphic endemics, restricted to one rock type such as Mancos Shale or gypsum. With regard to broader geographic patterns of plant endemism, the Colorado Plateau's "sky island" habitats are far richer in the number of endemic plants per unit of land area than its more extensive lowlands below 2000 m elevation (Welsh 1989). The La Sal, Abajo, and Henry Mountains, as well as Navajo Mountain, contain 11 rare and endemic plants not found elsewhere. However, because of the greater area of the lands below 2000 m on the Colorado Plateau, lower elevations harbor a greater number of endemics than higher elevations overall (Utah Division of Natural Resources 1998).

Another key element affecting geographic patterns of endemism is the peculiar distribution of hanging gardens, an aquatic habitat occurring at springs and seeps associated with near-vertical rock walls that is relatively unique to the plateau (Spence and Henderson 1993; Welsh 1989). At least 10 plants associated with hanging gardens are both endemic and rare, including the hanging gardens sullivantia, Navajo sedge, and the bog alcove orchid.

> The results of our study suggest that the ecoregion representing the highest rate of endemism with continental North America north of Mexico, at least in terms of actual numbers of species, is the Colorado Plateau Shrublands. This ecoregion, which spans more than 326,390 sq km of broad desert plains, interspersed by volcanic mountains and traversed by the Colorado River, contains more than 290 endemic species ... Due to its large physical size and varied topography, it is reasonable to expect that this ecoregion would be dotted by many small pockets of unusual habitat with numerous isolated plant populations, representing both paleoendemic and neoendemic species.
>
> —John Kartesz and Amy Farstad (1999)

Table 6. Estimated levels of species richness for the Colorado Plateau, relative to North America (the United States and Canada).

U.S. Species	Colorado Plateau
Vertebrates	
Mammals (1)	418
Birds (1)	776
Reptiles (1)	278
Amphibians (1)	242
Freshwater fish (1)	822
Vertebrate total (1)	2536
Invertebrates	
Butterflies/skippers (1)	600
Dragonflies/damselflies (1-2)	453
Tiger beetles (1-2)	111
Selected invertebrate total	1163
Vascular plants	
Ferns	546
Conifers	115
Flowering plants & trees	15,447
Vascular plant total	16,108

Armstrong (1982, 2002; Appendix 1) has suggested that 23.6 percent of all mammals on the plateau and 36 percent of all rodents exhibit endemism at the species or subspecies level. Eighty-three of these mammalian taxa occur on Indian reservations on the plateau, and 82 occur in national parks and monuments. Five mammals are found on reservations but not in parks and monuments, and another five are recorded in parks and monuments but not on reservations. High numbers of narrowly restricted species in invertebrate groups such as tiger beetles and leafhoppers are also suspected for the Colorado Plateau (Nelson 1994; L. Stevens and N. Cobb, personal comm. 2002).

Few biologists recognize that most of the endemic species on the Colorado Plateau are in genera which have been widely utilized by Native Americans as food, fiber, medicine, and ceremonial plants. Roughly two-thirds of the region's endemic plants (188 taxa, see Appendix 2) have been historically used or managed by the Navajo, Hopi, Hualapai, Havasupai, Laguna, and Acoma. (This inventory of 188 useful plant taxa on the plateau has been drawn from ethnographies too numerous to mention here, but which are being included in an electronic database in development at NAU's Laboratory of Ethnoecology and Indigenous Mapping.) Their insights into these plants' distributions and habitats could greatly help botanists who are attempting to know the status of these plants in Indian country (Nabhan 2000b).

At another level of diversity lie the exceptional populations and outstanding individual specimens within wide-ranging species of economic or ecological importance. At least 19 tree species of the Intermountain West have their most exceptionally large individuals still growing on the Colorado Plateau (Table 7); these "big trees" may be unique genetic variants, or simply individuals that have been able to thrive under favorable conditions for growth over many decades. Further inquiry is needed to determine whether they are unique genetic resources of this ecoregion.

Trends in Diminishing Biodiversity

From state, federal, and reservation-wide assessments of threatened species, we are aware of at least 70 high-priority (G-1) globally threatened plants on the Colorado Plateau, and another 60 second-priority plant taxa (G-2), as categorized by the Nature Conservancy and Center for Plant Conservation. Of these, at least 33 are now considered to be federally threatened or endangered (Table 8).

Povilitis (2000) ranked the threats that specifically imperil wildlife (endangered animals and plants) on the Colorado Plateau as follows: land development, livestock, water development, agriculture, resource extraction, hunting and collecting, off-road vehicle use, pollution and pesticides, direct disturbance, climate change, introduction of non-native species, and natural fire suppression. There is growing concern that some threats, such as competition from invasive species, may be increasing in impact relative to others. At least 155 exotic plant species are known to have naturalized in the Grand Canyon subregion of the plateau alone, and

Table 7. Big trees of the Colorado Plateau.

Species and Common Name	State	Circumference (inches)	Height (feet)	Crown (feet)
Abies concolor (White Fir)	UT	227	94	48
Abies lasiocarpa var. *arizonica* (Corkbark Fir)	NM	157	95	33
Alnus oblongifolia (Arizona Alder)	NM	199	129	50
Arctostaphylos pringlei (Pringle Manzanita)	AZ	37	18	14
Cupressus arizonica var. *glabra* (Ariz. Smooth Cypress)	AZ	196	70	44
Fraxinus anomala (Singleleaf Ash)	CO	19	24	23
Fraxinus anomala var. *lowellii* (Lowell Ash)	AZ	28	28	9
Juniperus deppeana (Alligator Juniper)	AZ	328	46	49
Juniperus monosperma (Oneseed Juniper)	NM	168	29	28
Juniperus scopulorum (Rocky Mountain Juniper)	UT	247	40	21
Ostrya knowltonii (Knowlton Hop Hornbeam)	AZ	39	44	29
Picea pungens (Blue Spruce)	UT	186	122	36
Pinus edulis (Two-leaf Pinyon)	NM	213	69	52
Pinus flexilis (Limber Pine)	UT	275	58	46
Pinus leiophylla (Chihuahua Pine)	AZ	121	87	34
Prunus serotina var. *rufula* (Southwest Black Cherry)	AZ	102	45	26
Quercus gambelii (Gambel Oak)	NM	216	47	85
Quercus grisea (Grey Oak)	NM	216	45	73
Salix gooddingii (Goodding Willow)	NM	354	45	89

now comprise 10.4 percent of that area's total flora (Stevens and Ayres, in press); a few of them—such as tamarisk and camelthorn—are dominant species in otherwise biologically diverse habitats in the region, such as spring and wetland settings. Exotic plant species richness appears to be highest in wetland, riparian, spring and seep, and North Rim ecosystems; the highest densities of exotics are at springs and seeps (15 species per sq km), with exotics in riparian corridors occurring at a tenth of that density, and other native habitats with far lower densities of non-native species (Stevens and Ayres, in press).

There are currently only five invertebrate species that are federally listed as rare or endangered on the Colorado Plateau or nearby landscapes: *Boloria acrocnema*—Uncompahgre fritillary butterfly (E); *Oxyloma haydeni kanabensis*—Kanab ambersnail (E); *Thermosphaeroma thermophilus*—Socorro isopod (E); *Valvata utahensis*—Utah snail (E); and *Cicindilia limbata albissima*—Coral Pink Sand Dunes tiger beetle (E). It is likely that other invertebrates are just as rare as these. At least nine exotic invertebrates now compete with natives on the plateau.

Nineteen taxa of vertebrates occur on the Colorado Plateau that are federally listed as threatened or endangered (Table 9); many more species are on "watch lists." For mammals, most of the endemic taxa listed in Appendix 1 should continue to receive conservation concern, particularly as exotic species alter their habitats (Armstrong 1982). Of particular concern among other vertebrates are native fish (Sigler and Sigler 1994). Of the original eight native fish species in the mainstream Colorado River and its tributaries in the Grand Canyon, only four remain, and they have met with predation and competition from 24 additional alien species of fish detected in the river corridor since 1963. Thus 85.7 percent of the fish of the regulated portion of the watershed are non-native. Three species of herpetofauna, four species of birds, and seven alien mammals (including feral livestock) also impact the native fauna and flora in innumerable ways (Stevens and Ayres, in press).

Despite the view of the World Wildlife

Table 8. Threatened and endangered plants on the Colorado Plateau.

Scientific Name	Common Name	Status
Arctomecon humilis	Dwarf bear-poppy	E
Asclepias welshii	Welsh's milkweed	T
Astragalus cremnophylax var. *cremnophylax*	Sentry milkvetch	E
Astragalus humillimus	Mancos milkvetch	E
Astragalus montii	Heliotrope milkvetch	T
Astragalus ousterhoutii	Ousterhout milkvetch	E
Carex specuicola	Navajo sedge	*T*
Cycladenia jonesii	Jones cycladenia	T
Erigeron maguerei	Maguire's daisy	T
Erigeron rhizomatus	Zuni fleabane	T
Erigonum gypsophilum	Gypsum wild-buckwheat	T
Hedeoma todsenii	Todsen's pennyroyal	E
Ipomopsis sancti-spiritus	Holy Ghost ipomopsis	E
Lepidium barnebyanum	Barneby ridgecress	E
Lesquerella congesta	Duelley Bluffs bladderpod	T
Lepidium tumulosa	Kodachrome bladderpod	E
Pediocactus bradyi	Brady pincushion cactus	E
Pediocactus despainii	San Rafael cactus	E
Pediocactus knowltonii	Knowlton cactus	E
Pediocactus peeblesianus var. *peeblesianus*	Peebles Navajo cactus	E
Pediacactus sileri	Siler pincushion cactus	T
Pediocactus winkleri	Winkler cactus	T
Phacelia argillacea	Clay phacelia	E
Purshia subintegra	Arizona cliffrose	E
Ranunculus aestivalis	Autumn buttercup	E
Senecio franciscanus	San Francisco Peaks groundsel	T
Schoenocrambe argillacea	Clay reed-mustard	T
Schoenocrambe barnebyi	Barneby reed-mustard	T
Schoenocrambe suffretescens	Shrubby reed-mustard	E
Sclerocactus glaucus	Vinta Basin hookless cactus	T
Sclerocactus mesaverdae	Mesaverde cactus	T
Sclerocactus wrightiae	Wright fishhook cactus	E
Spiranthes diluvialis	Ute Ladies' tresses	T
Townsendia aprica	Last chance townsendia	T

Fund and the Center for Plant Conservation that the status of the plateau's biodiversity is "relatively stable" (Ricketts et al. 1999b), changes in land and water use, the accelerated spread of exotic species, and a century of inappropriate grazing practices continue to threaten many populations, species, habitats, and landscapes in the region. Since 1900, at least 17 species of plants and animals have been extirpated from Grand Canyon National Park alone, despite the fact that this protected area has had a larger resource protection budget than most parks and reservations on the Colorado Plateau.

Prospects, Options, and Opportunities

The emergence of strong interagency task forces to monitor and protect biodiversity on the plateau is one of the most hopeful signs of the last decade. First, the five biennial conferences already held on research on the Colorado Plateau sponsored by the Colora-

Table 9. Threatened and endangered vertebrates of the Colorado Plateau.

Scientific Name	Common Name	Status
Canis lupus	Gray wolf	E
Epidonax traillii preblei	Southwestern willow flycatcher	E
Epoeciliopsis occidentalis	Gila topminnow	E
Falco peregrinus anatum	American peregrine falcon	T
Gila cypha	Humpback chub	E
Gila elegans	Bonytail chub	E
Gila robusta semidnuda	Virgin River chub	E
Gopherus agassizii	Desert tortoise	T
Grus americana	Whooping crane	E
Haliaeetus leucocephalus	Bald eagle	T
Lepidomeda vittata	Little Colorado River spinedace	T
Mustela nigripes	Black-footed ferret	E
Oncorhynchus apache	Apache trout	T
Oncorhynchus gilae	Gila trout	E
Plagopterus argentissimus	Wound fin	E
Ptychocheilus lucius	Colorado River pilceminnow	E
Strix occidentalis lucida	Mexican spotted owl	T
Ursus arctus	Grizzly bear	T
Xyrauchen texanus	Razorback sucker	E

do Plateau Field Station and Northern Arizona University have allowed ecologists to identify regional trends and to propose appropriate activities to better safeguard biodiversity. In particular, the interagency task force on invasive species control, nicknamed "Pulling Together," has had a major role in training teams to eradicate invasive species and to prevent others from getting a foothold in the region.

Similarly the Ecological Restoration Institute at NAU has spearheaded a regional effort to restore the ponderosa pine forests of the Colorado Plateau, reducing fire danger to neighboring communities at the same time. It remains unclear whether we will witness a dramatic increase in understory diversity within restored pine forests, or whether the effects of restoration on biodiversity will be more site specific, with considerable time lags in species composition change.

Unfortunately, the third of the plateau under Native American stewardship has to date been less affected by these interagency collaborative efforts. If all of the threatened species and habitats found on Indian lands were also found on lands rigorously protected by the National Park Service or the Nature Conservancy, perhaps this issue would be easier for conservation biologists to ignore. However, the narrowly distributed endemics of the region are often restricted to habitats found only on reservation lands, and not on park lands. For instance, the Navajo sedge is a sensitive species found around hanging gardens and seeps used by Diné livestock herders, and its range is largely restricted to the Navajo Reservation where they manage its habitat (D. House, personal communication; Nabhan et al. 1991). Similar conditions exist for the endemic Hopi chipmunk (*Tamias rufus*), a subspecies of the spotted ground squirrel (*Spermophilus spilosoma cryptospilotus*); the Chuska and Tunitcha Mountain subspecies of Abert's squirrel (*Sciurus aberti chuskensis*); and a subspecies of Stephen's woodrat (*Neo-*

toma stephensi relicta) found only on Navajo lands.

Fortunately, there has been tremendous growth in the size and technical quality of reservation-based Natural Resources, Forestry, Fish and Game, and Natural Heritage programs within the last two decades. Although the number of highly trained biologists and natural resource managers on reservations still lags far below that found in national parks, wildlife refuges, and the Nature Conservancy areas of comparable size, their ranks are swelling. In particular, efforts at the Zuni, Navajo, Hopi, and White Mountain Apache Reservations indicate that quality monitoring, restoration, and recovery programs are becoming more common.

One unresolved issue remaining is that of data sharing among biologists both on and off reservations. Until comparable surveys of biodiversity have been conducted both on and off reservations, and until there is a willingness to collaborate on habitat restoration and species recovery efforts that span the borders of sovereign nations, our capacity to act will be limited, and we will find it nearly impossible to even visualize an ecoregional conservation plan for linking protected areas through restored corridors. Although Mexican wolf and California condor reintroductions have proceeded, they will continue to suffer additional setbacks unless reservation communities become full participants.

ETHNOLINGUISTIC DIVERSITY: RESTORING THE NATIVE LANGUAGES TO THE COLORADO PLATEAU

Patrick Pynes and Gary Paul Nabhan

When traveling across the region, casual observers hardly conversant with cultural or biological geography nevertheless sense that they are in the midst of diversity, perhaps because of the spectacular ways in which the Colorado Plateau combines so many different habitats. Even more intriguing are the geographic juxtapositions that make these differences bold and explicit.

The Colorado Plateau's ecological combinations are truly remarkable, but biodiversity and geology are not the only dimensions of the ecoregion's unique and special character. Its human diversity is just as rich and remarkable as its diverse plants, soils, animals, microclimates, and ecological communities. Indeed, its linguistic and cultural diversity is a vital dimension of its distinct character, and its cultural landscapes are inhabited by people who have been living on the Colorado Plateau for hundreds or thousands of years. When Francisco Vasquez de Coronado's expedition of 1540 made Europe's first foray into Zuni territory in the area now called the Colorado Plateau, the linguistic and cultural diversity he encountered was palpable. For hundreds of miles in all directions surrounding the Zuni villages were numerous Apache, Havasupai, Hopi, Hualapai, Navajo, Paiute, Pueblo, Ute, and Yavapai settlements where people were speaking different, often mutually unintelligible languages, and practicing distinct cultural traditions.

Less than 50 years after Columbus's ships sailed into the Caribbean, the first large, organized group of Europeans entered the Colorado Plateau and greater Southwest at Zuni. Spanish-speaking people eventually brought with them everything from wheat, apple trees, and horses, to Castellano as a language unto itself. They also brought along smallpox and measles, which eventually decimated many indigenous populations. Agents of the Latin American mestizaje—genetically mixed, Spanish-speaking people—added to the Colorado Plateau's biological, linguistic, and cultural diversity in some ways, and subtracted from this diversity in other ways.

Zuni Pueblo's place in cultural exchange is even more intriguing when we consider that the Zuni language is an isolate; it has no known relatives in any other language family. Similar to how many rare and endangered plant species and communities on the Colorado Plateau are "endemic" to specific soils, microclimates, and ecosystems, the Zuni language is endemic to the southeastern portion of the Colorado Plateau.

Despite the fact that their language has no known relatives, the Zuni have been an integral part of the Puebloan cultures of the Colorado Plateau; a set of common cultural values is shared by most Pueblo people who live on the Colorado Plateau (Acoma, Hano, Hopi, Jemez, Laguna, Zia, and Zuni), or just east of it in the Rio Grande Valley (Taos, San Juan, Sandia, Isleta, Santa Clara). Contemporary Puebloan peoples are presumably descendants of the Anasazi—Navajo for "enemy ancestors"—an ancient amalgam of people whose advanced civilization rose and fell on the Colorado Plateau only a few centuries before the Coronado expedition.

Among the contemporary Puebloans of the Colorado Plateau—the Acoma, Hopi, Hopi Tewa, Towa (Jemez), Laguna, Zia, and Zuni—four distinct indigenous language families are represented. Besides Zuni, these include Keresan, dialects of which are spoken at Acoma Pueblo, Laguna Pueblo, and Zia Pueblo, and Kiowa-Tanoan, the language family to which the Towa (Jemez Pueblo) and Tewa (Hano Pueblo) languages belong (Figure 5). The Tewa-speaking people of Hano, who fled Spanish colonization in New Mexico during historic times and migrated to the Hopi mesas, may be the most endangered language community on the plateau. These Tewa-speaking Hopis are surrounded by Hopi speakers, who are surrounded by Navajo speakers, who are themselves surrounded by English and Spanish speaking communities. Hano's particular dialect of Tewa is no longer spoken in New Mexico.

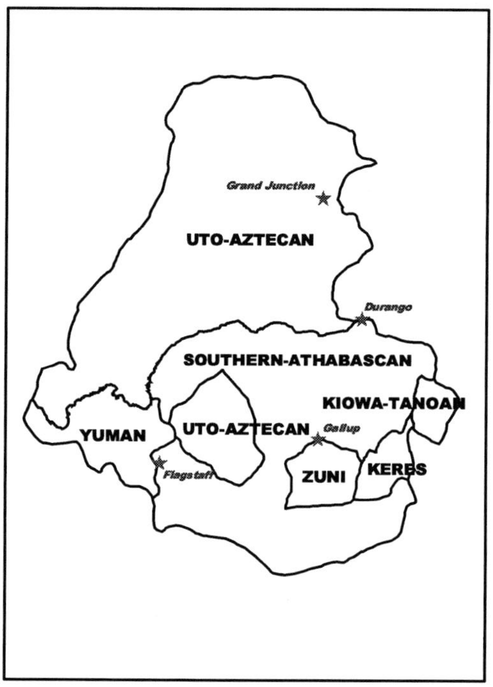

Figure 5. Language families currently represented on the Colorado Plateau and their historic distributions (map by Conservation Technologies Institute).

The fourth such indigenous language family is Uto-Aztecan, the family to which the Hopi language belongs (see Table 10). Hopi is the only Uto-Aztecan language spoken by any historic or contemporary Puebloans, however it is not the only Uto-Aztecan language spoken on the Colorado Plateau. The Ute and Paiute peoples of the western and northern portions of the Colorado Plateau also speak languages that belong to the Uto-Aztecan family. In 1540, Ute-speaking people were living in the parts of the Colorado Plateau that are now western Colorado, eastern Utah, and northwestern New Mexico, while Paiute-speaking people occupied mainly present-day northern Arizona and southern Utah. Today, after 150 years of U.S. governance, several Ute and Paiute tribes and communities are living on several different reservations within the states of Colorado, Utah, and Arizona. These reservations are severely reduced portions of roughly the same areas that Utes and Paiutes occupied and used before 1848, when the United States seized "El Norte" from Mexico.

Along with Zuni, Keres, Kiowa-Tanoan, and Uto-Aztecan, two other indigenous language families are currently represented on the Colorado—Yuman and Athabascan. The Yuman-speaking people of the Colorado Plateau include the Yavapai, Havasupai, and Hualapai of present-day Arizona. Their traditional homelands on the Colorado Plateau stretch from the Mogollon Rim, which marks the plateau's southern edge, to the north rim of the Grand Canyon and beyond. Today, the Yavapai are living on tiny reservations in central Arizona, and the Havasupai possess a small reservation mostly below the south rim of the Grand Canyon. The Hualapai are living on a somewhat larger reservation near the Colorado River, also along the Grand Canyon's south rim. During the past 100 years, much of the original homelands and traditional use areas of these Pai groups on both sides of the Colorado River has been usurped or surrounded by

Table 10. Status of indigenous languages on the Colorado Plateau.

Tribe, Band, or Community	Total Landbase (acres)	Language or Subgroup	Language Family Affiliation	Total Members	% Fluent Speakers	% Fluent Speakers Aged 2–18
Jicarilla Apache	879,605	Apache	Eyak-Athabaskan	3,500	23	<2
Tonto Apache	85	Apache	Eyak-Athabaskan	110	30	0
White Mountain Apache	1,600,000	Apache	Eyak-Athabaskan	15,000	50	13
Alamo Navajo	63,000	Apache	Eyak-Athabaskan	2,000	95	90
Navajo Nation	17,000,000	Navajo	Eyak-Athabaskan	259,556	57	<50
Ramah Navajo	146,953	Navajo	Eyak-Athabaskan	2,463	60	5
Tóhajiileehé (Cañoncito Band of Navajos)	80,000	Navajo	Eyak-Athabaskan	2,382	75	50
Havasupai	188,077	Pai	Cochimi-Yuman	639	98	90
Hualapai	1,000,000	Pai	Cochimi-Yuman	2,100	30	<25
Camp Verde Yavapai-Apache	636	Pai and Apache	Cochimi-Yuman & Eyak-Athabaskan	1,675	<2	<2
Yavapai-Prescott	1,395	Pai	Cochimi-Yuman	158	<2	<2
Kaibab Band of Paiute Indians	120,431	Southern Paiute	Uto-Aztecan	240	<2	0
Paiute Tribe of Utah	36,000	Southern Paiute	Uto-Aztecan	753	<2	0
San Juan Southern Paiute	5,000	Southern Paiute	Uto-Aztecan	300	N/A	N/A
Southern Ute	313,288	Ute	Uto-Aztecan	1,316	15	5
Uintah Ouray Ute	4,500,000	Ute	Uto-Aztecan	3,500	33	<33
Ute Mountain Ute	606,218	Ute	Uto-Aztecan	2,000	54	2
Acoma Pueblo	378,262	Western Keres	Keresan	6,344	50	2
Laguna Pueblo	491,387	Western Keres	Keresan	7,696	30	2
Hopi Pueblo	1,542,306	Hopi	Uto-Aztecan	10,916	48	<8
Hano (Hopi Tewa)	N/A	Tewa	Kiowa-Tanoan	600	65	<2
Jemez Pueblo	89,619	Towa	Kiowa-Tanoan	3,131	42	N/A
Zia Pueblo	117,000	Eastern Keres	Keresan	773	70	N/A
Zuni Pueblo	463,270	Zuni	Zuni (no known relatives)	9,690	66	60
24 others*	29,622,532	10 languages	6 language families	335,782	N/A	N/A

*24 distinct tribes, bands, or communities in four states.

To compile the data in this table, the following people were interviewed in person, via electronic mail, or by telephone (interviewers were Patrick Pynes, Tony Joe, and David Seibert): G. Benson, Paiute Tribe of Utah, telephone interview, December 2000. W. Eriacho, Bilingual Specialist, Zuni Public Schools, Zuni Pueblo, New Mexico, personal interview, November 2000. E. Gallegos, Culture and Language Department, Ute Mountain Ute Tribe, telephone interview, November 2000. Greg Glassco, Cultural Resources Office, Hualapai Tribe, Peach Springs, Arizona, telephone interview, December 12, 2000. J. Hussey, Assistant Director, Cedar City Paiute Tribe of Utah, Language Department, telephone interview, December 2001. D. Naranjo, Cultural Preservation Office, Southern Ute Tribe, Ignacio, Colorado, telephone interview, December 2000. J. Pino, Middle and High School Teacher, Alamo Navajo Community, Alamo, New Mexico, telephone interview, November 2000. M. Polacca, Office of Hopi Tribal Enrollment, telephone interview, December 2000. Anita Poleahla, Office of Hopi Cultural Preservation, telephone interview, December 10, 2000. G. Rollo, Tribal Administrator, Paiute Tribe of Utah, telephone interview, September 2000. N. Salari, Program Coordinator, San Juan Southern Paiute Tribe, telephone interview, September 2000. M. Sandoval, Cultural Preservation Department, Jicarilla Apache Nation, Dulce, New Mexico, telephone interview, November 2000. V. Saunders, Tribal Secretary, Tonto Apache Tribe, Payson, Arizona, telephone interview, September 2000. T. Secatero, President of Tohajilee (Canoncito Navajo), telephone interview, October 2000. C. Sims, electronic mail correspondence and personal interview, Acoma Pueblo, January 2001. J. Smith, Administrative Assistant, Camp Verde Yavapai-Apache Tribe, telephone interview, September 2000. N. Tohtsoni, Vital Affairs Office, Whiteriver, Arizona, White Mountain Apache Nation, telephone interview November 13, 2000. Also http:/ / primenet.com (landbase information) and "Who Is a Navajo?" Navajo Times. E. Wemytewa, Zuni Department of Game and Fish, telephone interview, September 2000. L. Willis, Director of Cultural Center, Jicarilla Apache Nation, Dulce, New Mexico, telephone interview, November 2000. Lyle McNeal, interview via electronic mail, October 2000.

the National Park Service at the Grand Canyon and other federal agencies.

The Athabascan-speaking people of the Colorado Plateau include the Navajo and Apache, whom many historians and anthropologists believe were the last indigenous people to settle on the Colorado Plateau before the arrival of Europeans. The Navajo and Apache are closely related linguistically; they speak separate but more or less mutually intelligible Athabascan languages. However, their cultures are distinctly different from one another. These cultural and linguistic differences were likely less pronounced at the time of the Coronado expedition: early Spanish explorers called the Southern Athabascans the "Apache Navajos" or "Navajo Apaches" (Hendricks and Wilson 1996). Today, there are important linguistic, cultural, and political differences among and between the Colorado Plateau's several different Apache and Navajo tribes, bands, and communities. Languages in the Athabascan language family are still being spoken across much of contemporary North America, from interior Alaska and the Yukon (Dene) to the American Southwest (Diné).

In fact, many of the people who are still speaking an indigenous language within the United States today are speaking Navajo (Diné bizaad: literally, "the People, their words"). According to the National Clearinghouse for Bilingual Education, there are 148,530 fluent Navajo language speakers in the United States today. Before 1492, at least 300 languages were being spoken in what is now the contiguous United States. Today, about 154 different indigenous languages are being spoken in the United States, nearly half of those original languages. There are 361,978 fluent speakers of these remaining 154 indigenous languages. Of these, more than half (184,504 or 51 percent) speak languages that are indigenous to the Colorado Plateau. Including Navajo, 5 of the top 20 remaining Native American languages in the United States are spoken by Colorado Plateau tribes: Navajo, Western Apache, Zuni, Hopi, and Keresan (in descending order, according to total number of fluent speakers; Estes 1999). English, Spanish, and Basque are also spoken in communities on the plateau.

> I can disappear into the [Navajo] forest and emerge somewhere else, someplace where I can believe it is a hundred years ago or even further back. Except for the contrails overhead, there is nothing to show the year. The forest is timeless ... The smell of damp earth is heavy in the canyons, ripe with the smell of rotting wood and leaves ... Some of the trees at the bottoms of canyons are giants, four feet in diameter, over one hundred feet tall, and hundreds of years old. There, in the absolute stillness, I can believe there is no Kit Carson. No Niña, Pinta, or Santa Maria.
>
> —Irvin Morris (1997)

Considering all of these contemporary statistics as well as the historical record, it is clear that the Colorado Plateau has been and continues to be an important center for the remaining indigenous linguistic and cultural diversity in the United States. One might say that the plateau is the primary repository or living sanctuary for these remaining indigenous languages. As the U.S. military discovered during World War II with the tremendous success of the Navajo and other Native American "code talkers," indigenous languages are one of this continent's most valuable treasures. They are vital and irreplaceable cultural resources for the Colorado Plateau, for the United States, and for the entire world.

> I believe that only in diné bizaad, the Navajo language, which is endless, can this place be described, or even indicated in its true character. Just there is the center of an intricate geology, a whole and unique landscape which includes Utah, Colorado, Arizona, and New Mexico. The most brilliant colors in the earth are there, I believe, and the most beautiful and extraordinary land forms—and surely the coldest, clearest air, which is run through with pure light.
>
> —N. Scott Momaday (1976)

Indigenous languages encode the knowledge, values, and relationships that native communities have accumulated, tested and nurtured in this place for thousands of years. This knowledge is often embedded in stories and ceremonies that have been passed down for many generations, carried along by living human voices, speaking across the generations. These voices do not normally speak English; to reach the full depth of meaning, they speak Apache, Havasupai, Hualapai, Hopi, Jemez, Keres, Navajo, Paiute, Tewa, Ute, Yavapai, or Zuni.

> This land that may seem arid and forlorn to the newcomer is full of stories which hold the spirits of the people, those who live here today and those who lived centuries and other worlds ago.
>
> —Luci Tapahonso (1993)

The Vitality of Indigenous Languages

Many different indicators show that the Colorado Plateau is an important "hotspot" for North America's remaining ethnolinguistic diversity, and this fact should be acknowledged and celebrated. However, a closer look at the present status of the plateau's indigenous languages reveals that this heritage is increasingly threatened. Some of the region's linguistic and cultural traditions remain strong and resilient, while others are being revitalized (Yamamoto et al. 2002); however, there are no guarantees that their languages will continue to be spoken in the future, especially if present trends continue. For although the Colorado Plateau is characterized by remarkable indigenous language diversity, there are increasingly visible signs that it can no longer be considered a place of indigenous language vitality.

Since World War II, and especially since the late 1950s, the region's indigenous language vitality has been gradually eroded. During this process of erosion, most of the plateau's Indian communities have been involved in a dramatic shift toward English as the primary language of daily discourse. In this sense, the English language itself is a major threat to the Colorado Plateau's linguistic and cultural diversity, particularly as English bombards communities via radio, television, and video.

Although the root causes for this rapid shift to English are beyond the scope of this report, several historic, economic, social, and political factors have been involved in this process. According to Chris Sims (personal communication 2001) of Acoma Pueblo, the erosion of the plateau's indigenous languages began when the federal government began forcing or coercing young Native American children to attend boarding schools far from home, and to reject their own languages in favor of the English language and the values associated with it on the frontier. English language-only education in the boarding schools was a main thrust of the federal government's official policy of American Indian assimilation. Soon, young Native Americans were surrounded by an entirely English speaking media, and there seemed to be few if any economic or social incentives for speaking an indigenous language.

Today, the dramatic effects of indigenous language loss are clearly visible across the United States, and these effects are becoming increasingly obvious on the Colorado Plateau. As recently as 1996 informed observers were still asserting that "only in the Southwest are many Native American languages relatively viable and vital" (Krauss 1996). However, here at the beginning of the twenty-first century, all of the Colorado Plateau's indigenous languages—including Navajo—have lost much of their former vitality, and some may even be threatened with loss of entire domains in their historic lexicon that dealt with traditional ceremonial and subsistence activities no longer practiced.

Most young Native Americans (ages 2–18) who live in the Colorado Plateau's indigenous communities or urban centers are growing up as monolingual English speakers. Although some children may understand a few words or phrases in their native tongue, most do not speak their tribal languages fluently or conversationally. This is the way things are across the entire ecoregion, including Hopi Pueblo, which many people consider the most "traditional" of the Colorado Plateau's indigenous communities. A recent survey and study of Hopi language use by Daniel Higgins (2001) found that although 100 percent of Hopis aged 60 or older are fluent speakers of the Hopi language, only about 8 percent of Hopi children aged 2–18 can speak the language fluently. If trends continue, in another generation, far less than a quarter of all Hopis will be conversant in their mother tongue.

According to Sims (personal communication 2001), the generational shift to English at Acoma and Laguna Pueblos has been even more advanced than the shift at Hopi. The "shift to English ... within the last two decades has been so dramatic ... at Acoma and Laguna [that] one can pretty safely say that no child entering preschool is speaking Keres as their first language any more." She added that, "by and large you will not hear school age children or older adolescents in these communities [Laguna and Acoma] using Keres as a medium of communication in their daily home or social lives. It has all shifted to English" (Sims, personal communication 2001).

Several other recent formal surveys of language use in the Colorado Plateau's indigenous communities have begun to draw more precise and detailed pictures of what is happening as the overall shift to American English continues (Yamamoto et al. 2002). The majority of these surveys have been conducted by the tribes themselves, with some financial assistance from the federal government's ANA (Association for Native Americans) program and other sources. In some ways the pictures that these surveys are drawing seem bleak. Taken as a whole, they confirm the pattern: the majority of the Colorado Plateau's indigenous young people are growing up speaking English as their only language.

There are, of course, significant differences in language status among the plateau's tribal communities, stemming from the linguistic and cultural differences that distinguish the tribes from one another, and from each community's unique history and distinct experience of colonialism. As shown in Table 10, our own research and that of Yamamoto and colleagues (2002) illustrates these different histories of language retention, and current efforts at language revitalization. Two of the plateau's most geographically isolated indigenous communities, the Alamo Navajos and the Havasupais, have the highest percentages of young people who can speak their own language fluently (90% for each). On the other hand, very few young people in the Ute, Paiute, Pueblo, and Apache communities on the plateau can speak their own indigenous language fluently (generally 2% or less). Overall, only 8 of the 24 indigenous communities on the Colorado Plateau report that at least 8 percent of their young people can speak their own language fluently. If these figures are accurate, then only a small minority of contemporary Native American youths on the Colorado Plateau are growing up speaking the same language that their ancestors spoke.

For Navajo, Hopi, Basque, or Zuni, the vitality of any human language is created and maintained by older generations orally passing on the language to younger generations. When the crucial connections between the generations are disrupted or broken, a language's vitality inevitably erodes. In a worst-case scenario, if present patterns continue, some or all of the plateau's indigenous languages could become moribund, or even extinct by the end of the twenty-first century. For people who speak one or more of the Colorado Plateau's indigenous languages, and for people who understand and care about the beauty of the plateau's indigenous languages, this scenario is unthinkable and unacceptable, especially given the vital link between place, identity, and language.

Alternative Scenarios and Opportunities for Language Revitalization

Despite declines, there are reasons to be hopeful and optimistic that these homogenizing trends can and will be reversed. Just as restoration ecologists believe that they can restore the plateau's ponderosa pine forests, linguists and cultural educators also believe that they can work with and assist indigenous language communities as they try to restore the vitality of their own languages (Yamamoto et al. 2002). As the Hopi Language Summit in 2001 demonstrated, everyone must "win" to make a language persist within a community: classroom teachers and teacher's aides, elders, language scholars, singers, storytellers, and, most important, students.

In fact, numerous language revitalization efforts are already underway in several of the Colorado Plateau's indigenous communities. Within the past few years, virtually every plateau tribe has established or has planned cultural resource and language departments. Organized and supported by their respective tribal governments, these departments are actively assessing the status of their own languages and are making efforts to stabilize or restore their vitality (Yamamoto et al. 2002).

Of the Colorado Plateau's 24 tribal communities, perhaps the Navajo Nation's language revitalization efforts are the most comprehensive. This is not surprising, given that the Navajo Nation is the largest Indian tribe in the United States, in both total landbase and population. At present, their efforts include Navajo language courses in both tribally controlled and non-tribally controlled schools (K–12), community-based Navajo language immersion camps for both young people and adults, language teacher training programs, and college or university level Navajo language courses. To one degree or another, all of the tribal governments on the plateau are engaged in similar language education efforts, although few have sufficient resources to offer the language education opportunities that the Navajo Nation offers (see Table 11).

The capacity of the plateau's indigenous communities and governments to become actively involved in such efforts is increasing, but more help is needed. Consider, for example, the Paiute Indian Tribe of Utah. Although tribal membership totals only 720 people, and a 1998 ANA-sponsored survey determined that only 3 percent of all tribal members (all of them elders) were fluent Paiute speakers, the community has not given up hope that its language can be restored. Based on data from the 1998 survey, the tribe's language department designed a second ANA grant to "encompass a preschool, a tribal elder teaching component, and a home study program" (Hussey, personal communication 2001).

Today, the tribe operates a language immersion preschool for tribal members aged 3–5 who live in the Cedar City area. A tribal elder has been trained to teach the Paiute language to the children, using songs and stories. This language immersion preschool is a 9-month program designed to complement the tribe's "Project: First Language." This project is "designed for the parents of babies, first-time parents, or preschool students. This is to help teach traditional parenting such as songs, stories, crafts, language, and traditional theories ... The program is funded by ANA ... unfortunately, the future is contingent on outside funding" (Hussey, personal communication 2001).

Recommendations for Future Action

The language revitalization experts whom we surveyed recommended investment in certain elements they think are essential to any plateau tribes' efforts to maintain and restore their languages. Some of these elements can be gleaned from the experience of the Paiute Indian Tribe of Utah's revitalization efforts. One is that a crucial first step in revitalization efforts is assessing the status of the speaking community of a specific language. Like the Paiute Indian Tribe of

Table 11. Present status of tribal efforts to protect and revitalize the indigenous languages of the Colorado Plateau.

Tribe, Band, or Community	K–12 Tribal Schools[1]	K–12 Nontribal Schools[2]	Language Immersion Youths[3]	Language Education Adults[4]	Teacher Education Programs[5]	College or University Courses[6]
Jicarilla Apache	–	x	x	–	–	–
Tonto Apache	–	–	–	–	–	–
White Mtn. Apache	–	–	–	–	–	–
Alamo Navajo	–	–	–	–	–	–
Navajo Nation	x	x	x	x	x	x
Ramah Navajo	–	–	–	–	–	–
Tóhajiileehé (Cañoncito Navajo)	–	x	–	–	–	–
Havasupai	x	–	x	–	–	–
Hualapai	–	x	x	–	–	–
Camp Verde Yavapai-Apache	x	–	x	–	–	–
Yavapai-Prescott	–	–	x	x	–	–
Kaibab Band of Paiute Indians	–	–	x	x	–	–
Paiute Tribe of Utah	–	–	x	–	–	–
San Juan Southern Paiute	–	–	–	–	–	–
Southern Ute	x	x	x	–	–	–
Uintah Ouray Ute	x	x	–	–	–	x
Ute Mountain Ute	–	–	x	x	–	–
Acoma Pueblo	x	–	x	x	–	–
Hano (Hopi Tewa)	–	–	–	–	–	–
Laguna Pueblo	–	–	–	–	–	–
Hopi Pueblo	–	–	–	–	–	–
Jemez Pueblo	–	–	–	–	–	–
Zia Pueblo	–	x	x	x	–	–
Zuni Pueblo	x	x	x	x	x	–

1. K–12 language courses taught in tribally controlled schools.
2. K–12 language courses taught in non-tribally controlled schools.
3. Community-based language immersion camps for youth.
4. Community-based language education programs for adults.
5. Language teacher education and training programs available.
6. College or university level language courses available.

Utah, several of the plateau's indigenous communities have used ANA grants to support these kinds of surveys and other initial language revitalization efforts. Besides the federal government, many tribes have also received funding from nonprofit foundations on the plateau, namely the Lannan Foundation and the Indigenous Language Institute (ILI), both based in Santa Fe, New Mexico. ILI focuses a great deal of its attention upon the Southwest, including the Colorado Plateau. Among many other efforts, ILI offers tribes technical support for their language revitalization efforts, and looks for ways to bring together and synergize the efforts of various tribes and like-minded nonprofit organizations.

> The Indigenous Language Institute recognizes the imminent loss of indigenous peoples' languages and acknowledges the individuality of indigenous communities. ILI facilitates innovative, successful community-based initiatives for language revitalization through collaboration with other appropriate groups and organizations, and promotes public awareness of this crisis.
>
> —ILI Mission Statement (1998)

Federal and nonprofit funding sources have been important in helping the tribes to get the language revitalization process underway. However, because federal grants typically fund only one or two years of work and research, it has been difficult for the tribes to build long-term momentum, as the Paiute Indian Tribe of Utah has discovered ("the future is contingent upon outside funding"). Because the revitalization process will require years and even decades to "complete," patience and profound changes in attitudes will be required.

Along with long-term support, direct and active community involvement is another important theme in indigenous language revitalization efforts across the Colorado Plateau. For these projects to actually achieve something tangible, people of all ages within the speaking communities must be willing to participate in them, as did the tribal elder who got involved in teaching Paiute children their language through songs and stories. Unless community members support restoration efforts themselves and get directly involved in them, no amount of outside intervention to stabilize and revitalize the languages can be successful. Ultimately, native people must do it themselves. And "Doing it yourself" is what tribal sovereignty is all about.

> Our languages are for our people. We have the right to decide if we even want to save our language. We are the only ones who can save our language.
>
> —Marci Naranjo
> Southern Ute tribal elder
> (pers. comm. 2000)

After building a foundation for cross-generational collaboration, another important step in the restoration process is to provide education and training for those who are teaching tribal members how to speak their own language. The Paiute elder who got involved in teaching young children first received education and training; the elder did not just show up in the classroom, thinking that his knowledge of the language would automatically make him a good teacher. The necessity for educating language teachers is a theme that we encountered in talking to tribal language department personnel across the plateau. Wilfred Eriacho, Sr., Director of Bilingual Education for Zuni Pueblo's Public School District, postponed retirement to help lead efforts to train Zuni language teachers, affirming that "we [Zunis] have to homegrow our Zuni teachers" (personal communication 2000).

One of the best approaches to language teaching is a "natural" immersion process in the home and community where young children are raised within a specific indigenous language environment that creates and defines their cultural habitat. Language immersion camps and projects attempt to recreate these kinds of intense cultural contexts, especially for youths.

The message is clear: Although some of the Colorado Plateau's languages are declining in health and vitality just as rapidly as some of our natural resources, many communities are experimenting with language education and revitalization projects, and they are evaluating the most effective ways to slow down, stop, or reverse the erosion of their languages and related cultural traditions.

One of the most creative of these recent efforts is Hopi community radio station KUYI, which broadcasts from Hotevilla on the Hopi Nation at 88.1 FM. Many of KUYI's disc jockeys speak Hopi over the airways, and are heard up to 60 miles from Hotevilla, among Navajo, Southern Paiute, Anglo, and Hispanic communities as well. In its constant shifting back and forth between Hopi and English, between rock and roll, reggae, blues, Mexican conjunto, country-and-western, and traditional American Indian music, KUYI embodies the Colorado Plateau's remarkable cultural diversity, and the station models innovative ways of bringing endangered indigenous languages into novel contexts, helping to reconnect young people with the beauty, vitality, and relevance of their language and therefore their culture.

In December of 2001, the first annual Yuman Language Summit, held on the Hualapai Indian Reservation, was aimed at "keeping our [Yuman speakers] languages alive," and included panels for both elders and youths, as well as roundtable discussions between scholars and practitioners. Important information, experiences, and strategies were shared; the summit is another good example of how indigenous people are "doing it themselves." In June of 2001, more than 50 people from nine cultures came together to discuss ways to "Bridge Ecological Restoration and Language Revitalization Efforts in Native American Communities." The consensus view of this workshop's participants was that ecological and language revitalization efforts have many goals in common, and many of the same tools, which are best used as part of community-based projects (a summary of retreat outcomes is available at http://www.terralingua.org). Perhaps the future of indigenous language revitalization efforts on the Colorado Plateau will be more closely intertwined with ecological restoration efforts, with participants using the plateau's diverse habitats as outdoor classrooms, places where people can learn about the vital relationships between the ecoregion's languages, cultures, and physical landscapes. It is up to all of us to create and nurture these linkages, and to assist communities on the Colorado Plateau to heal, evolve, and to continue, as Dan Simplicio of Zuni put it, not to merely preserve a culture, but to live it.

Agricultural Diversity: Crop Genetic Resources, Agrohabitats, and the Farmland–Wildland Mosaic on the Colorado Plateau

Gary Paul Nabhan and Patrick Pynes

> Native Americans in the Southwestern United States have farmed successfully in precarious arid and semiarid environments for over a millennium. Some of the methods used by the Zuni include the careful placement of fields on alluvial fans, complex manipulation of runoff, and management of gully formation. It appears that harvesting water and sediment from drainages has allowed the Zuni to favorably influence soil moisture, nutrient status, and texture of soil within their fields.
>
> —Roman R. Pawluk (Pawluk et al. 1992)

For several millennia, indigenous farmers of the Colorado Plateau have been cultivating species derived from their own regional biota, as well as adapting domesticated species from other regions to their homelands. Some of the oldest agricultural remains in North America occur not far from the plateau's southeastern periphery, at Bat Cave on the Plains of San Augustin in central New Mexico. Although the most ancient dates for the remains of cultivated plants found at Bat Cave are still being debated, it appears that these agricultural artifacts are no less than 3500 years old (Ford 1981).

These archaeobotanical materials give us a rough indication of how long the resident cultures of the Colorado Plateau have been influencing local biodiversity by intentionally managing useful plants and animals (Minnis and Elisens 2000). They also provide a historical baseline for estimating the time required for these crops to diversify and adapt to the region's unique range of agroecological conditions (Bingham and Bingham 1979; Woosley 1980). In turn, these conditions are influenced by a staggering diversity of soil types, hydrological conditions, and microclimates, which native farmers selected and modified for the best results. For instance, White and colleagues (1996) have demonstrated that crops grown in the cobble mulch gardens of prehistoric Anasazi cultivators produce four times greater yields than bare control plots on the same soil types.

As a result of the historical time depth over which Native Americans have farmed in this region, its broad environmental amplitude, and the large variety of cultural influences on the domesticated plants and animals tended there, the Colorado Plateau continues to be a center of agroecological diversity, harboring possibly the greatest species richness of native agricultural crops in the Americas north of the Tropic of Cancer (Winter 1974; Nabhan 1985).

Many of the crop species in the region are heirloom varieties and land races. Because some varieties of agaves, amaranths, sunflowers, and tepary beans appear to have originated only on the Colorado Plateau (Table 12), some geographers have considered the ecoregion a secondary center for crop origins and diversification (Nabhan 1985). In other words, a single crop species (such as maize or common beans) has diversified into a great number of varieties with distinctive agronomic characteristics, flavors, and uses, each contributing to the

Table 12. Native crops known from the Colorado Plateau before Europeans arrived (Nabhan 1985).

Scientific Name	Common Name	AZ	UT	NM	CO	Unique Varieties
Agave murpheyi	Hohokam agave	E	–	–	–	X
Agave utahensis	Anasazi agave	E	–	–	–	X
Amaranthus cruentus	Red grain amaranth	X	–	0	–	X
Capsicum annuum	Chile pepper	0	0	0	0	X
Canavalia ensiformis	Jack bean	E	–	E	–	–
Cucurbita argyrosperma	Striped cushaw	X	–	X	–	–
Cucurbita moschata	Big cheese pumpkin	X	–	X	–	–
Cucurbita pepo	Acorn squash	X	X	X	X	–
Gossypium hirsutum	Cotton	E	E	E	–	–
Helianthus annuus	Sunflower	X	–	0	–	X
Lagenaria siceraria	Bottlegourd	X	E	X	–	X
Nicotiana attenuata	Pueblo tobacco	X	X	X	–	X
Nicotiana tabacum	Common tobacco	X	X	X	X	–
Nicotiana rustica	Turkish tobacco	E	–	–	–	–
Opuntia viridiflora	Pueblo cholla	–	–	E	–	X
Physalis philadelphica	Zuni tomatillo	–	–	X	–	–
Proboscidea parviflora	Devil's claw	X	X	–	–	–
Phaseolus acutifolius	Tepary bean	0	0	0	–	–
Phaseolus coccineus	Runner bean	E	–	E	–	–
Phaseolus lunatus	Lima bean	X	–	–	–	X
Phaseolus vulgaris	Common bean	0	0	0	0	X
Zea mays	Corn, maize	0	0	0	0	X

E, Now extinct among indigenous communities, although possibly represented in seed banks.
0, Severe genetic erosion or varietal loss.
X means that this crop has been grown by a specific group, but the number of varieties is unknown. A dash indicates that the crop is not known to have been part of a specific group's agricultural traditions.

genetic diversity of the region's native farms. For example, the Hopi language names at least 15 folk varieties of maize, and Hopi farmers have been growing these varieties for centuries. Corn geneticists would classify these folk varieties into at least five groupings of land races (Whiting 1942). Although we currently lack a comprehensive list of all folk varieties or land races of native crops found among pre-Columbian farming communities on the Colorado Plateau, our best estimate is more than 75 land races. Native people of the plateau domesticated turkeys, macaws, and dogs before contact with Europeans, and the integration of turkeys into native agriculture and ceremonialism was unique to the Colorado Plateau (McKusick 1986). For domesticated crops, the genetic diversity among flour and flint maizes, sunflowers, grain amaranths, lima, tepary, and common beans (Table 13) is especially unique compared to the crop diversity found in adjacent ecoregions (Nabhan 1985).

After Spanish colonial influences entered the region, native and Indo-Hispanic farmers adopted additional sets of livestock and garden, field, and orchard crops. Brought to Mexico and the Southwest by the Spanish, the Navajo Churro sheep (whose origins are in North Africa) became the oldest adapted livestock breed to persist in a single North American ecoregion (Table 14; Christman et al. 1997). Criollo corriente cattle and Spanish mustangs were shared over a broader geographic range, but continue to be raised on the plateau. In addition, Jesuit and Franciscan priests introduced a number of vegetable and fruit tree varieties from Europe, Mexico, Africa, and Asia to the Colorado Plateau, and they continue to persist in the region (Table 15). These heirloom cultivated

Table 13. Indigenous agroecological diversity: One dozen common crop varieties grown by Colorado Plateau tribes (adapted from Joseph Winter's unpublished dissertation, Department of Anthropology, University of Utah, 1974).

	Apache	Hopi	Navajo	Pai	Pueblo	S. Paiute
Corn (maize)	3–5	15	7–11	4–7	5–8	1–4
Common beans	2–3	5–12	6	2–6	4–12	1–3
Tepary beans	–	4	X	2–3	2–3	X
Lima beans	–	6	–	X	–	–
Pepo squash	X	–	3	X	5	–
Mixta squash	–	X	X	X	3	X
Moschata squash	–	X	–	–	–	–
Unspecified squash	2–3	–	10	–	X	–
Cotton	–	X	–	–	X	–
Gourds	X	X	X	X	2–4	X
Sunflowers	X	X	X	X	X	X
Amaranth	–	X	–	–	X	–

Note: Specific numbers (3–5, etc.) refer to the number of varieties grown by each group. X means that this crop has been grown by a specific group, but the number of varieties is unknown. A dash indicates that the crop is not known to have been part of a specific group's agricultural traditions.

Apache = Jicarilla, White Mountain, and Tonto, in New Mexico and Arizona; Hopi = Hopi Pueblos and Hano Pueblo, in Arizona; Navajo = all Navajo-speaking tribes, bands, and groups, including Ramah and other outliers, in Arizona, New Mexico, and Utah; Pai = Havasupai, Hualapai, and Yavapai, in Arizona; Pueblo = New Mexico Pueblos, including Acoma, Jemez, Laguna, Zia, and Zuni only; Southern Paiute = Shivwits, Kaibab, and San Juan, in Utah and Arizona. Although historically the Ute tribes of Colorado and Utah did engage in some limited agriculture, these groups have not been included here because of a lack of complete and reliable data.

plants added to the already existing crop diversity of the plateau without much initial loss of indigenous crop genetic diversity.

Along with domesticated plants and animals, at least two other domains of plants contribute to the Colorado Plateau's crop genetic resources. The first domain is that of wild relatives of crops, or species within the same botanical genera as the domesticated plants themselves. For the Colorado Plateau, these wild species include agaves, amaranths, beans, cucurbits, potatoes, prickly pear and cholla cacti, sunflowers, tobaccos, and tomatillos. Some of these species, such as wild potatoes (*Solanum fendleri* and *S. jamesii*) are harvested by several of the plateau's native communities. Navajo, Hopi, and Zuni gatherers collected these wild potatoes from both natural and agricultural habitats, eating them with clay. Native people used other wild plants, like buffalo gourd, as medicine and for oil. More recently, however, modern crop geneticists have domesticated buffalo gourd to make use of its root starches and oil-seeds.

A second domain of horticultural genetic resources on the Colorado Plateau includes useful wild species that are genetically unrelated to any domesticated crops, but that served foragers in the past and have new potential to serve as sources of industrial products and nursery material for landscaping and reclamation, or as representatives of the growth potential and potential longevity of their species. One example of a wild crop that has been used for industrial purposes is jojoba (*Simmondsia chinensis*). This plant's genetic variation has been collected, hybridized, and brought into cultivation, first on the southern reaches of the plateau, and then in a dozen other countries. Jojoba's unique product is a liquid wax with proven value in maintaining the health of hair follicles, now used in more than two dozen commercially available shampoos. A native plant that has been used in landscaping and reclamation projects is the Arizona cypress (*Cupressus arizonica*), which was brought into cultivation by the Soil Conservation Service during the Dust Bowl of the 1930s, and is now

Table 14. Rare historic livestock breeds persisting on the Colorado Plateau.

Common Name	Scientific Name	Comments
Merriam's wild turkeys	*Meleagris gallopavo merriami*	Naturalized domesticated turkeys of prehistoric Anasazi, Sinagua, and Mimbres cultures, now common (in the tens of thousands) in ponderosa pine and pinyon-juniper woodlands.
Black Spanish turkeys	*Meleagris gallopavo*	Critically endangered domesticated breed developed by Mexican and Western European farmers, now rare in the Southwest.
Navajo churro sheep	*Ovis aries*	First domestic sheep breed in the New World (1540 AD), adapted to desertscrub of the arid plateau, once endangered but now undergoing a revival, with roughly 50,000 head now in production, primariliy among Navajo and Hispanic weavers.
Criollo corriente cattle	*Bos bos*	First resident cattle breed in the U.S. Southwest (1690 AD), adapted to desertscrub and pine-oak woodlands; once threatened, but now being revived for rodeos and roping competitions.
Spanish mustang/ Spanish barb	*Equus caballus*	Established in the U.S. Southwest from Spanish introductions by 1500 AD, mustangs were selected by Native Americans for their hardiness and smaller sizes. Several mustang lines, including the Wilbur-Cruz population from southern Arizona, remain in the region.
Feral burros	*Equus asnus*	Burros introduced into the U.S. Southwest by 1500 AD have persisted in feral populations and have interbred with later introductions.

Primary Source: Christman et al. 1997.

propagated from Arizona genetic stock in more than two dozen countries. Colorado Plateau trees that serve as the benchmarks for their species' potential growth and longevity were listed in Table 7 (page 45).

> Dependence on agriculture does not necessarily mean independence of wild gathered foods ... Gathered foods retained their importance [although] floods and drought would have affected cactus, beeweed and pinyon throughout the area and would have imperiled the subsistence system, regardless of what was happening in the cornfields.
>
> —Earl Morris, quoted in Stoker (1990)

Trends

Unfortunately, these gene pools of useful plants and animals were not valued or conserved to the extent that they should have been during most of the twentieth century. They have suffered from four different pressures, or threats: (1) the wholesale abandonment of traditional agriculture for economic or other reasons, or the loss of the lands and waters that had once made these traditions possible; (2) the replacement of native crops with modern hybrid crops of the same or different species; (3) the accidental hybridization and "swamping-out" of heirloom varieties by modern varieties; and (4) the loss of traditional knowledge regarding seed saving, selection, and propagation, leading to genetic erosion (Nabhan 1985). As a result of these four factors, many of the plateau's Native American and Indo-Hispano crops and domesticated animals suffered dramatic reductions in their gene pool variation and sizes. Navajo Churro sheep, for example, nearly became extinct, with fewer than 435 animals after World War II (McNeal, personal communication, October 2000).

The loss of individual crops or breeds is not the only biocultural impoverishment that has occurred in the region over the last

Table 15. Field and orchard crops introduced by the Spanish for which there are unique heirloom varieties on the Colorado Plateau.

Scientific Name	Common Name
Allium cepa	Onion
Allium	Garlic
Citrullus lanatus	Watermelon
Coriandrum sativum	Coriander
Cucumis melo	Melon
Cucurbita maxima	Hubbard squash
Malus	Apple
Mentha	Mint
Opuntia ficus-indica	Prickly pear
Prunus	Peach
Prunus	Apricot
Prunus	Plum
Vicia faba	Fava
Vitis vinifera	Grape

half century. When families living on farms dropped to 2 percent of the total U.S. population in the mid 1980s, only about 3 percent of the Native American population—20,000 families—persisted with farming in any form. Probably only a small percentage of these farmers grew the heirloom crops of their ancestors (Nabhan 1986). By the mid 1980s, most of the fields and orchards that had been cultivated as native agrohabitats for centuries had been abandoned on the Colorado Plateau. Thus, not only the seeds and breeds were being lost, but their unique habitats also vanished from the landscape. In a very real sense, these trends continue.

> All agree that it is crucial to promote not only research but also grassroots education, accompanied by policy recommendations that are morally compelling ... To this end, the scientific database will be most effectively used if joined with concepts that citizens of the United States already cherish. The first is love of the land.
>
> —E. O. Wilson (Stein et al. 2000)

Opportunities

These losses in agricultural biodiversity did not go unnoticed by traditional farmers or crop geneticists. Between 1948 and 1951, the U.S. National Research Council attempted to collect many of the land races of Colorado Plateau maize for ex situ conservation, but many of their unique values were lost when grown in Iowa, where there was counter-selection for their adaptations to Colorado Plateau conditions. Then, in the 1980s, Native Seeds/SEARCH, a nonprofit organization based in Tucson, Arizona, with board and staff personnel including members of the Acoma Pueblo, Hopi, and Navajo Tribes, was successful in returning earlier-collected germplasm samples to the tribes of Arizona, New Mexico, Utah, Colorado, and Sonora, Mexico, thus bridging ex situ with in situ conservation efforts. Since then, several other traditional agricultural projects, including tribal efforts based on Zuni, Hopi, Navajo, and San Juan Pueblo lands have been moderately successful in restoring the value of traditional agriculture in the minds of community members. In 1994, the Traditional Native American Farmers Association had its founding meeting in Gallup, New Mexico, and most participants were from tribes residing on the Colorado Plateau. Other nonprofit organizations, including the Talavaya Project, the original Seeds of Change project, and Navajo Family Farms have made valuable contributions to these endeavors, but have been less successful in finding ways to survive financially while still ensuring that local communities receive tangible benefits from ecoregional seed conservation efforts.

On the other hand, the Navajo Sheep Project (NSP) and the Navajo Churro Sheep Association, both nonprofit organizations cofounded by Churro sheep expert Dr. Lyle McNeal, along with a Navajo organization Diné Be'iina (Sheep Is Life) and the American Minor Breeds Conservancy (AMBC), have had great success in rescuing the gene pools of rare sheep and turkey breeds to benefit contemporary livestock keepers (Christman et al. 1997). According to McNeal (personal communication, October 2000), at least 3000 churros are currently living on the Navajo Reservation. AMBC has received the Slow Foods Ark award for its work with these rare livestock breeds.

Following these successes is a more recent effort to protect and restore the Colorado Plateau's historic agroecological habitats. At Hopi, the Bacavi agricultural terraces have been restored and revived through a collaboration between the Hopi Cultural Preservation Office and Northern Arizona University's Applied Anthropology Program. The Second Mesa villages have designated six Hopi fields as "Good Farms" to serve as models for arable land protection near the mesa. A similar protection and restoration effort is now underway for spring-fed apricot and peach orchards at Sipaulovi.

> Most recently, the [Bacavi terrace] project has expanded beyond restoration and documentation to hands-on educational activities to involve the younger children of the community. As one [Hopi] mother commented, "If they are going to learn the traditional Hopi way, then the children need to have the terrace gardens to work with."
>
> —Miguel Vasquez and
> Leigh Jenkins Kuwanwisima
> (1994)

Peach orchards have been revived on Navajo lands in Canyon de Chelly as well. There are plans underway for the agricultural restoration of fields, orchards, and pastures at Hubbell Trading Post National Historic Site in the Ganado Lake Watershed on the Navajo Reservation. Zuni Pueblo's Sustainable Agriculture Project has been successful in reviving many of the tribe's traditional agricultural crops and practices, and several other New Mexico Pueblos have undertaken pilot projects to do the same.

Scenarios and Options for the Future

From the mid 1990s onward, there has been frequent collaboration among farmers from various tribes in renovating their fields and orchards, sharing seed stock, and marketing their produce as "Indian grown." Non-Indians have also taken up growing regionally adapted heirloom seeds, producing crops largely for farmer's markets, gourmet restaurants, and community-supported agriculture coalitions. Should the landscape mosaic of small-scale agriculture and restored riparian gallery forests ever emerge again, one might expect an increase in breeding bird densities similar to those that were found historically on the Upper Rio Grande. As Steve Emslie (1981) has documented for the New Mexican pueblos, there was an apparent decline in avian species richness along northern New Mexico floodplains when small-scale agriculture and riparian forests were replaced by large dams, large fields, and urban development. We recommend restoring both wildlands and farmlands to their former diversity; we assume that if a healthy interplay between wildlands and farmlands is re-established, the recovery of the avifauna will follow.

Based on the Colorado Plateau's extant historic heirloom crops, rare livestock breeds, and improving prospects for upland and riparian forest restoration efforts, it seems quite possible that contemporary agriculture on the Colorado Plateau can be redesigned to build upon rather than to negate this region's native agroecological diversity. If indigenous agricultural traditions can persist anywhere in North America, we believe that it will be on the Colorado Plateau.

DIVERSIFYING CONSERVATION STRATEGIES: INNOVATIVE MEANS OF PROTECTING TRADITIONAL CULTURAL PLACES AND THEIR WILDLANDS SETTINGS

Tony Joe, Gary Paul Nabhan, and Patrick Pynes

The parks, monuments, and other protected areas of the Colorado Plateau have received international recognition for nearly a century for exemplifying some of America's pioneering efforts to formally conserve natural and cultural landscapes (Sellars 1977). National parks and monuments, wilderness areas, primitive areas, wildlife refuges, world heritage sites, and other federal protected areas were established in or near the Colorado Plateau soon after each designation was formalized as a conservation strategy. These protected areas, in addition to state parks and "private" Nature Conservancy areas, are typically what form the core of any ecoregional plan for conserving land, water, and biodiversity (Soulé and Terborgh 1999). Nevertheless, the resource managers and conservationists involved in establishing the first federally protected areas on the plateau had little scientific data on which to base their efforts or to delineate boundaries for parks and refuges. It would have been impossible for early planners of protected areas to have a deep understanding of how biocultural diversity has shaped what they deem precious and worthy of conserving within this ecoregion.

Since the establishment of Mesa Verde National Park in 1906—followed by Grand Canyon National Park in 1919 and Canyon de Chelly in 1930 (Burnham 2000)—more than 10.8 million acres of the Colorado Plateau's 130 million acres have been federally protected for their natural and cultural resources (Table 16). In retrospect, some of these protected areas have been established more or less in the right places but for the wrong reasons, at least from the perspective of traditional communities residing in or near them. As Burnham (2000) has recently demonstrated, the history of protected areas on the Colorado Plateau reads like a synopsis of the history of cross-cultural conflict about land in western North America—the history of whose lives are enriched and whose are sometimes impoverished by shifts in land tenure.

Once that is acknowledged, there is little doubt that the landforms, waters, and biota of the Colorado Plateau are considered to be of great scenic, recreational, cultural, and biological value to legions of local residents as well as tourists who come from diverse backgrounds. Although they may not be aware of the statistics that confirm how unique this ecoregion is, many visitors to these protected areas have a sense of the plateau's earthly and spiritual blessings. For many people, the Grand Canyon of the Colorado River embodies this uniqueness. To experience its grandeur first-hand, more than 5 million people visit Grand Canyon National Park each year; many of these visitors have been lured to the region by the beautiful images of the Colorado Plateau's canyons, mesas, rivers, mountains, and other spectacular landforms that have graced the covers of countless national and international magazines, and have been the

Table 16. Federally protected areas on the Colorado Plateau.

I. Wilderness Areas on the Colorado Plateau (NPS, BLM, USFS)

Arizona

Bear Wallow (Apache National Forest: 11, 080 acres)
Beaver Dam Mountains (BLM Arizona Strip District: 15,000 acres)
Blue Range Primitive Area (Apache-Sitgreaves National Forest: 173,762 acres)
Cottonwood Point (BLM Arizona Strip District: 6,860 acres)
Escudilla (Apache National Forest: 5,200 acres)
Fossil Springs (Coconino National Forest: 22,149 acres)
Grand Wash Cliffs (BLM Arizona Strip District: 37,030 acres)
Kachina Peaks (Coconino National Forest: 616 acres)
Kanab Creek (Kaibab National Forest & BLM Arizona Strip: 70,460 acres)
Kendrick Mountain (Coconino and Kaibab National Forests: 6,510 acres)
Mount Baldy (Apache National Forest: 7,079 acres)
Mount Logan (BLM Arizona Strip District: 14,650 acres)
Mount Trumbull (BLM Arizona Strip District: 7,880 acres)
Munds Mountain (Coconino National Forest: 24,411 acres)
Paiute (BLM Arizona Strip District: 87,900 acres)
Paria Canyon-Vermilion Cliffs (BLM Arizona Strip: 89,400 acres)
Petrified Forest (NPS [within Petrified Forest National Park]: 50,260 acres)
Red Rock-Secret Mountain (Coconino National Forest: 47,194 acres)
Saddle Mountain (Kaibab National Forest: 40,539 acres)
Strawberry Crater (Coconino National Forest: 10,743 acres)
Sycamore Canyon (Coconino, Kaibab, and Prescott National Forests: 55,937 acres)
West Clear Creek (Coconino National Forest: 15,238 acres)
Wet Beaver (Coconino National Forest: 6,155 acres)
Total Wilderness in Arizona portion of Colorado Plateau: 797,053 acres

Colorado

Black Canyon of the Gunnison (NPS [in Black Canyon of the Gunnison National Park]: 15,599 acres)
Black Ridge Canyons (BLM: 70,370 acres)
Gunnison Gorge (BLM: 17,700 acres)
Lizard Head (San Juan, Rio Grande, and Uncompaghre National Forests: 41,193 acres)
Mesa Verde (NPS [within Mesa Verde National Park]: 8,100 acres)
Mount Sneffels (Uncompaghre National Forest: 16,565 acres)
Total Wilderness in Colorado portion of Colorado Plateau: 132,527 acres

New Mexico

Apache Kid (Cibola National Forest: 44,626 acres)
Bisti/De-na-zin (BLM Albuquerque District: 59,450 acres)
Blue Range (Apache and Gila National Forests: 29,304 acres)
Cebolla (BLM Albuquerque District [within El Malpais National Conservation Area]: 62,800 acres)
Chama River Canyon (Santa Fe and Carson National Forests: 50,300 acres)
Cruces Basin (Carson National Forest: 18,000 acres)
San Pedro Parks (Santa Fe National Forest: 41,132 acres)
West Malpais (BLM Albuquerque Dist. [in El Malpais National Conservation Area]: 39,700 acres)
Withington (Cibola National Forest: 19,000 acres)
Total Wilderness in New Mexico portion of Colorado Plateau: 344,312 acres

Utah

Ashdown Gorge (Dixie National Forest: 7,043 acres)
Beaver Dam Mountains (BLM Cedar City District: 2,600 acres)
Black Ridge Canyons (BLM: 5,180 acres)
Box–Death Hollow (Dixie National Forest: 25,751 acres)
Dark Canyon (Manti–La Sal National Forest: 47,116 acres)
High Uintahs (Ashley and Wasatch National Forests: 456,705 acres)
Paria Canyon–Vermilion Cliffs (BLM Cedar City District: 20,000 acres)
Pine Valley Mountain (Dixie National Forest: 50,232 acres)
Total Wilderness in Utah portion of Colorado Plateau: 614,627 acres

Total acreage of Colorado Plateau Wilderness Areas: 1,888,519 acres (1.9 million acres)

Table 16 (continued)

II. National Parks on the Colorado Plateau (NPS)

Arizona
- Grand Canyon (1,217,403 acres)
- Petrified Forest (93,533 acres)

Colorado
- Black Canyon of the Gunnison (27,705 acres)
- Mesa Verde (52,121 acres)

Utah
- Arches (76,518 acres)
- Bryce (35,845 acres)
- Canyonlands (337,597 acres)
- Capitol Reef (241,904 acres)
- Zion (146,592 acres)

Total acreage of National Parks on Colorado Plateau: 2,229,218 acres (2.3 million acres)

III. National Historic Parks / Sites on the Colorado Plateau (NPS)

Arizona
- Hubbell Trading Post National Historic Site (160 acres)

New Mexico
- Chaco Culture National Historic Park (33,974 acres)

Total acreage of National Historic Parks / Sites on Colorado Plateau: 33,954 acres

IV. National Monuments on the Colorado Plateau (NPS and BLM)

Arizona
- Canyon de Chelly* (NPS: 83,840 acres)
- Grand Canyon–Parashant (BLM: 1,014,000 acres)
- Montezuma Castle (NPS: 857 acres)
- Navajo* (NPS: 360 acres)
- Pipe Spring (NPS: 40 acres, and a mule)
- Tuzigoot (NPS: 40 acres, and another mule)
- Sunset Crater (NPS: 3,040 acres)
- Vermilion Cliffs (BLM: 293,000 acres)
- Walnut Canyon (NPS: 3,579 acres)
- Wupatki (NPS: 35,422 acres)

Colorado
- Canyons of the Ancients (BLM: 164,000 acres)
- Colorado (NPS: 20,453 acres)
- Yucca House (NPS: 33 acres)

New Mexico
- Aztec Ruins (NPS: 319 acres)
- El Morro (NPS: 1,278 acres)
- El Malpais (NPS: 114,277 acres)

Utah
- Cedar Breaks (NPS: 6,154 acres)
- Dinosaur (NPS: 210,277 acres)
- Grand Staircase–Escalante (BLM: 1,900,000 acres)
- Hovenweep (NPS: 784 acres)
- Natural Bridges (NPS: 7,636 acres)
- Rainbow Bridge (NPS: 160 acres)

Total acreage of National Monuments on Colorado Plateau: 3,859,549 acres (3.8 million acres)
*Located on Navajo Nation land, but leased by the NPS.

Table 16 (continued)

V. National Recreation Areas on the Colorado Plateau (NPS)

Arizona
Glen Canyon (1,254,306 acres)
Lake Mead (1,495,665 acres)

Colorado
Curecanti (41,972 acres)

Total acreage of National Recreation Areas on Colorado Plateau: 2,791,943 acres (2.8 million acres)

VI. National Conservation Areas and National Preserves on the Colorado Plateau (BLM and USFS)

Colorado
Colorado Canyons National Conservation Area (BLM: 122,240 acres)
Gunnison Gorge National Conservation Area (BLM: 57,725 acres)

New Mexico
El Malpais National Conservation Area (BLM: 262,100 acres)
Valles Caldera National Preserve (USFS*: 90,000 acres)

Total acreage of National Conservation Areas/National Preserves on Colorado Plateau: 532,065 acres

*Although the United States Forest Service has legal title to the land, the nine-member Valles Caldera Trust (whose members are appointed by the U.S. President, in consultation with New Mexico's congressional delegation) will make management decisions for this National Preserve in the Jemez Mountains west of Los Alamos, New Mexico, on the far eastern edge of the Colorado Plateau. Two members of the Trust include the Supervisor for the Santa Fe National Forest and the Superintendent of Bandelier National Monument. President William Clinton signed legislation establishing this spectacular National Preserve in July 2000. Most recently a private working ranch, Valles Caldera National Preserve is not yet open to the public.

VII. National Wildlife Refuges on the Colorado Plateau

Colorado
Browns Park (13,455 acres: Maybell, Colorado)

Utah
Ouray (11,827 acres: Vernal, Utah)

Total acreage of National Wildlife Refuges on the Colorado Plateau: 25,282 acres

Grand total of federally protected lands on the Colorado Plateau (National Conservation Areas, National Historical Parks/Sites, National Monuments, National Parks, National Preserves, National Recreation Areas, National Wildlife Refuges, Wilderness Areas: 11,184,071 acres (11.2 million acres)

Federally protected lands located within larger federally protected land management units (i.e, wilderness areas within national parks) have not been counted twice.

Sources: Bureau of Land Management (http://www.blm.gov); National Wilderness Preservation System (http://www.wilderness.net/nwps); *Wilderness Areas of the Colorado Plateau* (Museum of Northern Arizona 1989); U.S. Fish and Wildlife Service (http://www.fws.gov); National Park Service (http://www.nps.gov).

subject of several influential books, from *Desert Solitaire* by Ed Abbey (1968), to *Blessed by Light: Visions of the Colorado Plateau*, edited by Stephen Trimble (1991).

We applaud past efforts to conserve land as parks and monuments on the Colorado Plateau. We also think that the notable accomplishments during the next century of conservation in this region will be not so much attracting European tourists or literary naturalists, as providing lasting quality-of-life benefits to the region's permanent residents. Conservation discussions in the region are already dominated not by what tourists or seasonal park rangers value in the Colorado Plateau, but rather what long-time residents value in their own homelands—a myriad of natural and cultural features that are collectively known as Traditional Cultural Properties or Places (TCPs). We prefer "Places" over "Properties" because "Properties" connotes non-indigenous concepts of individual land ownership, rather than stewardship rights and privileges held in common, with inherent obligations to past and future generations.

> The question of homogeneity and heterogeneity is crucial to both cultures and biological species. In the case of local communities living around or in parks, cultural integrity is at stake. With loss of biological diversity like that ... found within the walls of Canyon de Chelly National Monument, *Homo sapiens* is faced with the potential lack of genetic maneuverability in the future and loss of possible sources of medicines and crops.
>
> —Teresa N. Stoker (1990)

We suggest that as a supplementary strategy to parks, monuments, and refuges, the designation of Traditional Cultural Places is an underutilized but potentially effective means of celebrating, safeguarding, and restoring the relationships between resident people and the habitats that nurture them. In addition, it may be the only viable designation for protecting landscapes and water features on the Indian reservations, which comprise a third of the land area of the plateau, where "park" or "wilderness" designations are politically not an option because the loss of tribal lands to the National Park Service is still remembered so bitterly (Burnham 2000). Unlike other federal, state, or private land-protection designations—wilderness areas, primitive areas, conservancy areas, wildlife refuges, state parks, or national parks and monuments—the legal designation of TCPs as national historic landmarks or cultural landscapes does not function to evict or constrain the very cultural communities that have participated in shaping these landscapes for centuries.

The collective indigenous heritage and historic cultural features and places associated with Hispanic, Basque, and European American immigrant groups (including the Mormons) together comprise the predominant cultural traditions of the Colorado Plateau. Taking this cultural diversity into account, we explore what Traditional Cultural Places are, and propose why more formally designated TCPs may become vital to the plateau's living cultural communities. We also argue that the designation of TCPs should be more widely recognized as a conservation tool and be strengthened legally to provide fuller protection against vandalism, overuse, and encroachment. In addition, we describe some of the opportunities that exist for cross-cultural collaborations to identify, protect, and conserve these important places, making use of land stewardship designations that relatively few conservation organizations currently understand or utilize.

> We need, without the saying of it, to see the arrow-shaped deer tracks, fossils, shards of an older civilization, to read the language of an ancient past that is still being spoken in stone, by wind, in petroglyphs. We want to weave together the threads of where we come from with what we are now and where we are going. Our human journey here joins with the stories of snake, owl, lizard, the far-traveling winds and curving waters. We gauge our place by these stories, by our movement, and by our defining the fierce human struggle for survival, for whatever meaning and revelation live in this burning light of day.
>
> —Linda Hogan (1992)

Current Status and Trends

On the Colorado Plateau, more than 42,000 officially named places have been recorded in USGS databases, some of which echo the creation stories, history, ceremonial life, and surviving oral literature of the region's long-term inhabitants. Whether located on Indian reservations, in Mormon farm villages, or in Basque sheep-herding meadows, many of these place names reverberate with words first spoken in events that occurred long ago—Oraibi, Tuzigoot, Wupatki, Hovenweep, Hawikuh, Jemez, Pagosa, Lukachukai, Escalante, Hanksville, Lee's Ferry. Some places, such as the San Francisco Peaks, have been named in several indigenous languages and dialects. Towering more than 4000 m above sea level, and among the highest promontories seen from the Colorado Plateau's vast interior, "the Peaks" are considered sacred to so many different cultures that former Interior Secretary Bruce Babbitt negotiated the closure of an extractive pumice mine on their slopes. (In an ecumenical gesture, he also reminded his Native American colleagues that some European-Americans consider the Peaks sacred because of their presumed association with St. Francis of Assisi, the Patron Saint of Ecology.) The mine was purchased by the U.S. Department of the Interior and then closed.

> Where language touches the earth, there is the holy, there is the sacred. In our deepest intelligence we know this: that names and being are indivisible ... that the essential things of the world and the universe were and are in place, in place. They are fixed forever in their names.
>
> —N. Scott Momaday (1995)

Former Secretary of the Interior Babbitt is but one of an increasing number of federal and other government officials who acknowledge that the Apache, Havasupai, Hualapai, Hopi, Navajo, Paiute, Ute, Yavapai, Zuni, and other tribes have legal rights to maintain living spiritual connections to places like the Peaks, loosely guaranteed by the American Indian Religious Freedom Act (AIRFA) of 1978. These officials have also recognized that traditional cultural places on the Colorado Plateau can be legally protected through means in addition to AIRFA. Although several mountain ranges such as the San Francisco Peaks have multicultural significance, thousands of other places across the Colorado Plateau that are significant to only a single traditional community are nevertheless deemed essential to their identity as a people. Wherever a person may be standing on the Colorado Plateau, it is likely that he or she is within sight of one or more TCPs, whether designated or not. They literally surround us.

> The fact is, national parks and other designated preserves overlay—in their entireties—old homelands of traditional peoples ... not a square mile of the entire country lacks site-specific or contextual associations with the multiple millennia of Native Americans' cultural histories. That is why, in a new era of cultural awareness and inclusion, these multitudes of tribes and nations seek our welcome and sensitive accommodation as they continue, or renew, traditional ties to their old homelands.
>
> —William E. Brown (2001)

Besides entire mountain ranges or their prominent peaks, many of the most important indigenous TCPs on the Colorado Plateau are connected to the ecoregion's water—its springs, rivers, and lakes (Stoffle 1997). Many Navajo place names refer not only to the presence of perennial or ephemeral water, but also to local hydrological conditions, often in very specific ways. As a result, they can be used as benchmarks for documenting environmental change through time. Some of the Navajo places associated with water in the sacred Chuskas (the Navajo word *tó* = water) include Tsaile ("it flows into the rock," a perennial creek at the upper mouth of Canyon de Chelly), Tohatchi ("water is scratched out"), Todilto ("water that pops, sounding water"), Tó Dzís'á ("water that extends away into the distance," Wheatfields, Arizona), Tontsaa ("big water"), and Toadlena ("water flows up and out").

Indeed, many of the nearly 800 springs that can be found on USGS maps of Navajo,

Hopi, White Mountain Apache, and other tribal lands on the Colorado Plateau are regarded as sacred places, or are used to obtain minerals, mineral waters, plants, or feathers used in ceremonies. Many of these springs are suffering increasing environmental damage because the aquifers that supply them are being pumped and "mined" for industrial, agricultural, and other purposes. For many people, mining of source-pools such as the N-aquifer, which feeds sacred springs, is seen as a threat not merely to single habitats, but to a balanced hydrological cycle in its entirety. In response to Peabody Coal's use of N-aquifer water, Native American activists have printed thousands of bumper-stickers that read "Save NAVAH_2OPI—Stop Peabody Pumping." Both Hopi and Navajo activists have asked, Could the N-aquifer itself be designated as a TCP?

> Near Shiprock five horses stand at the left side of the road, watching traffic. A pole carrying talk cuts through the middle of the world. They notice the smoking destruction from the Four Corners plant as it veers overhead, shake their heads at the ways of these thoughtless humans, lope toward the vortex of circling sands where a pattern for survival is fiercely stated.
>
> —Joy Harjo (1989)

Traditional Cultural Places can be classified into several functional categories, as Richard Stoffle (1997) and his associates have done on behalf of the Southern Paiute. There are landmarks, some of which are important to the ecoregion's indigenous people because they are "sites of emergence" for specific tribes or clans. For example, the Navajo place of emergence is located in the San Juan Mountains of southwest Colorado. The Hopi place of emergence is located in the Grand Canyon. There are also waterscapes and landscapes that have ancient narratives attached to them, some of which are the routes for ritual pilgrimages. TCPs are therefore vitally important to the continued transmission of each tribe's traditional ecological knowledge of its homelands on the Colorado Plateau.

Unfortunately, the traditional ecological knowledge related to traditional cultural places on the Colorado Plateau is being increasingly threatened by the erosion of indigenous languages and a host of other threats. Protecting this connection means conserving history, culture, and language, as well as ensuring that the land or water feature itself is conserved. As Keith Basso (1996) has explained, TCPs are historical, ethical, and ecological touchstones, guiding tribes as they become more actively involved in protecting, conserving, and restoring their lands, languages, and cultures. Despite all of the losses that have already occurred, much has been saved, and the level of indigenous participation in land conservation, restoration, and stewardship efforts is increasing. As Andrew Gulliford (2000) has suggested, "Quietly but inexorably, Native Americans have become intimately involved in the preservation of their historic and sacred sites."

Many members of Colorado Plateau tribes with spiritual connections to the San Francisco Peaks, for example, played active and key roles in the Sierra Club's successful campaign to close the pumice mines on its slopes. Indigenous communities across the Intermountain West have also become involved in the Trust for Public Lands' successful Tribal Lands Program. These efforts have resulted in the return of important lands to the Nez Perce Tribe and to Santa Clara Pueblo.

And yet, it remains difficult for some people who are not conversant with indigenous cultures to understand why they might consider a mountain, river, or spring just as sacred as a cathedral, mosque, or synagogue. As documented in the film "In the Light of Reverence" (McLeod and Maynor 2001), tourists and local residents from other cultures may regard rivers, springs, mesas, mountains, and canyons as exceptionally beautiful, inspiring, or physically challenging, but they do not necessarily see those places as being intrinsically sacred. The cultural landscape—the sacred geography of a particular place—is invisible to them, and despite fairly direct wording in the American Indian Religious Freedom Act that the United States must protect access to places where religious traditions can be expressed, the law remains invisible in many courtrooms.

> Not once have the courts saved a sacred place exclusively on Native American arguments that their right to freely practice their religion would be compromised if the place were destroyed.
>
> —Steve Moore
> (Smith and Manning 1997)

Cultural differences have led many people to value "built" objects and environments over "natural" objects and environments, especially in terms of where they practice their religions, and with what talismans they guide their rituals. But are historic human constructions intrinsically more deserving of protection and restoration than places where sacredness was somehow "found" to be present before humans began constructing things there?

In contrast to this way of thinking, many indigenous people and other multigenerational residents value features that possess both natural and cultural dimensions simultaneously, like the cliff dwellings nestled within Canyon de Chelly, or Munkuntuweap, a weeping rock in Zion National Park where a Wolf spirit is said to reside (Milne 1995). Here a "natural" mountain range and sacred "cultural" landscape are completely intertwined. With regard to these biocultural landscapes, Gulliford has reminded us that "structures are secondary to sacred places" for tribal people. These fundamental cultural differences may explain why the ancient Anasazi Pueblo of Chaco Canyon is so highly regarded by the National Park Service—to the extent that Navajos who historically lived and camped near these ruins continue to be restricted from engaging in the traditional subsistence activities that they can legally undertake on National Park Service lands. Even with Chaco Canyon's more recent designation as a World Heritage Site by UNESCO, the National Park Service continues to have frequent conflicts with local Navajos, who painfully recall how their rights to lands within what is now Chaco Canyon National Historic Park were usurped by the NPS (Burnham 2000).

> The Park Service has done a very good job of preserving antiquities, but it has had trouble dealing with living cultures.
>
> —Jim Martin (Smith and Manning 1997)

Of course, many indigenous people have already experienced the destructive effects of having their connection to sacred places disregarded. The quarrying of Woodruff Butte near Holbrook Arizona—a TCP important to Hopi, Zuni, and Navajo people—is a tragic case in point (Linford 2000). When tribes objected to the Arizona State Historic Preservation Office, the federal and state governments' use of cinders from the privately controlled quarry on Woodruff Butte was suspended. However, the private mining firm that was leasing the quarry found other, nongovernmental users for the materials extracted there. As a result of these mining activities, shrines on Woodruff Butte have suffered significant damage, as documented in the film "In the Light of Reverence" (McLeod and Maynor 2001). A similar threat is now facing Zuni Salt Lake, a salt pilgrimage destination point for some tribes. A multicultural effort is attempting to prevent Arizona's Salt River Project from engaging in coal and aquifer mining activities near the lake. Although the lake has been part of the Zuni cultural landscape for centuries, like many indigenous TCPs it is not presently located on Zuni Reservation lands. The Zunis and their non-Zuni supporters believe that the proposed mining activities will damage the lake's spiritual and ecological values for generations to come.

> It's the most precious thing I can think of to know absolutely where you belong. There's a whole emotional wrapping-around-of-you here. You see the same rock, tree, road, clouds, sun—you develop a nice kind of intimacy with the world around you. To be intimate is to grow, to learn. To be intimate with emotional space, physical space, that is what is absolutely fulfilling. Intimacy, that's my magic word for why I live here.
>
> —Tessie Naranjo, Santa Clara Pueblo
> (personal communication 2000)

It is critical that all involved understand the current strengths and weaknesses of identifying traditional cultural places, especially if formal designations do not lead to more vigorous vigilance and on-the-ground protection. Both TCP and National Historic Landmark designations do not always tangibly result in both protection and appropriate cultural access to important places. However, they are initial steps that give recognition of the importance of a site to state, tribal, and federal authorities, who are then obligated to observe activities there and to be held responsible if protection remains inadequate. At that point, communities can often attract institutional support for additional protection, or for outright purchase of off-reservation sites. However, protecting TCPs is often more complicated than simply protecting endangered species or their habitats, apparently remote from cultural influences, and it is not itself a panacea.

> If [native peoples'] tenacity is the secret, then the secret inside it is the core value that creates the tenacity: a reverence—think that word through—for the land, for a particular place. Romanticism? The story of the Colorado Plateau makes it plain that, in this age when we careen toward an uncertain destination, a true and lasting commitment to place may be as valuable to us as any serum.
>
> —Charles Wilkinson (1999)

Prospects and Options

During the past century several federal laws as well as state and tribal statutes have been passed to protect the culturally significant resources associated with TCPs. These laws define precisely what TCPs are and how they can be protected, both by legal means and by grants to communities who can serve as stewards on the ground. Some of these laws include the Antiquities Act (1906), the National Historic Preservation Act (NHPA, 1966), the Archaeological and Historic Preservation Act (AHPA), the American Indian Religious Freedom Act (AIRFA, 1978), and the Archaeological Resources Protection Act (ARPA, 1979). In the process of tribal, state, and federal attempts to regulate access to TCPs and to build capacity to deal with their protection, several amendments have been added to these laws during the past 25 years. The purpose of these amendments has been to make the laws more operative and effective. Considered as a portfolio of strategies for cultural landscape preservation, these laws shape the process of designating and protecting a particular range of cultural resource areas, such as plant-gathering grounds, ceremonial sites, and other historic features associated with the broad conceptual and legal framework of traditional cultural places. However, because these laws have been conceived in somewhat piecemeal fashion, and are enforced by different regulatory agencies, there are still gaps in legal protection for TCPs.

In the cases of AIRFA and NHPA, places and properties have sometimes been narrowly defined within the Anglo-American legal frame of reference, so that all elements of what a native community considers significant and/or sacred have not necessarily been protected from damage or destruction. NHPA, in particular, has done a superb job of protecting and preserving important European-American historic sites on the Colorado Plateau, such as Winslow, Arizona's beautiful La Posada Hotel, designed by renowned architect Mary Colter, and built by the Santa Fe Railroad and Fred Harvey Company. However, the same law has done very little to protect the Colorado Plateau's indigenous TCPs, especially the ones that are not part of the "built" environment. As Gulliford (2000) wrote, "the legal framework for protecting Indian culture and cultural sites remains tenuous."

Nevertheless, as tribes gain sophistication in identifying and designating TCPs on and off reservations, they are utilizing more recent legal precedents that foster cross-cultural understanding about these issues and that best utilize a mix of existing legal frameworks as effective tools for protecting indigenous TCPs. Once identified, nominated, and designated into whatever category of legal recognition they fit, TCPs come under the protection of existing federal, state, and tribal authorities who are mandated to monitor and protect cultural resources. Each tribe on the Colorado Plateau has a different perspective and history of involvement with TCP designations.

The most important document to enter into this ongoing legal and cultural process was National Register Bulletin 38, prepared by Parker and King (1990), "Guidelines for Evaluating and Documenting Traditional Cultural Properties." This bulletin explains "Indian sacred site issues to non-Indians" and provides a legal framework for "identifying and evaluating the significance of such sites for National [Historic] Register listing" (Gulliford 2000). The bulletin legally defines TCPs as properties or places "that are eligible for inclusion in the National Register because of their association with cultural practices or beliefs of a living community that ... are rooted in that community's history, and ... are important in maintaining the continuing cultural identity of the community." Gulliford (2000) wrote that "the idea behind National Register Bulletin #38 was sound, reflecting an appreciation for Indian religious ideas and concepts of natural landforms as having deep meaning for tribal peoples who know such places through their origin stories, clan migrations, annual ceremonies, and traditions."

To further their legal protection, some TCPs have also been nominated as National Historic Landmarks (NHLs). They could also be nominated as National Natural Landmarks, which the Secretary of the Interior must then ratify should the Secretary concur with the National Park System Advisory Board that these cultural properties or places are of national significance. Since 1935, more than 2300 properties have been designated and protected as a result of federal agencies' collaborations with private owners, states, and other institutions, including many sites of significance to Native Americans (NPSAB/USDI 2001). The National Park Service also has management responsibility for the National Register of Historic Places, which includes sites of local or state significance, and the National Natural Landmarks program, which has been inactive the last few years. Because one of us (Nabhan) has served on the National Park System Advisory Board appointed by Congress, and has been involved in the review and designation of more than 50 National Historic Landmarks, it is clear to us that the current board and NHL staff are open to very broad, culturally sensitive designations (NPSAB/USDI 2001).

It is assumed by many that NHLs must be buildings or historic districts, but fishing grounds, hunting runs, fields, agricultural terraces, and plant gathering and processing sites are also admissible under the "agriculture/subsistence" function and use category, and rock art is admissible under the "recreation and culture" category. In addition, a "landscape" category can include plazas, gardens, forests, mountains, rivers, and lakes, and may even cover the lightning-struck trees sacred to the Navajo Enemy-Way traditions. In short, National Historic Landmarks clearly cover a range of cultural interactions with diverse natural as well as cultural habitats, and they ultimately offer more legal protection and operational maintenance or vigilance to communities wishing to be stewards of designated areas.

The NHL categories of significance cover "aboriginal" as well as non-aboriginal historic and religious sites equally, although sites currently used for religious purposes are not eligible unless they are deemed to be of national significance for architectural, artistic, or historic reasons. Nevertheless, sites can be designated if they are deemed significant to a particular ethnic heritage and if the community members associated with them are willing to tell the history of persons "having origins in any of the original peoples of North America, of Spanish-speaking peoples, or other ethnicities." Areas are also deemed significant and worthy of designation if they celebrate a culture's landscape architecture, science, religion, or social history in a unique manner. Even National Natural Landmarks (NNLs) can celebrate an individual's or culture's role in preserving the uniqueness of a particular habitat. For instance, an NNL designation could provide protection to the landscape where C. Hart

Merriam defined the "life zones" of the Colorado Plateau; NNLs could also be established to protect the springs used by Jemez Pueblo farmers to "prognosticate drought."

Conclusions and Recommendations

As Burnham (2000) and Nabhan (1997) have amply documented, Native Americans have not only been evicted from their homelands in what are now designated as National Parks and Wildlife Refuges, but they have often been excluded from conservation discussions and decisions as well. One historic reason for these exclusions is that cultural uses of natural habitats were thought to be incompatible with narrowly defined goals for the "preservation of nature"—as if cultural uses could only damage or destroy habitats assumed to be pristine wilderness. Today, however, if conservationists wish to foster broader collaborations in land stewardship on public and private lands, they must open themselves to working with Native American communities, whose reasons for wishing to see a site safeguarded from urban or industrial development may be very different from their own reasons. If designation of places as Traditional Cultural Properties, National Historic Landmarks, or National Natural Landmarks helps to protect their cultural significance, such designations may ultimately protect places as natural habitats for ecoregional biodiversity as well. Among the habitats currently designated as TCPs, NNLs, and NHLs are sacred peaks, springs, lightning-struck trees and snags, meadows, canyons, orchards, fields, and caves. Because many of these habitat types also harbor significant plant, bird, and mammal populations, their significance to biodiversity conservation is considerable. Protection as a TCP or NHL rather than as a National Monument or National Wildlife Refuge may be more acceptable cross culturally as well.

After consultations with tribal representatives from several indigenous nations and communities on the Colorado Plateau, the Center for Sustainable Environments proposes the following short-term goals:

1. Initiate a comparison of existing tribal efforts to inventory and designate TCPs and NHLs to discern how some tribes have advanced the use of these conservation tools. Then, hold training workshops for tribal professionals who wish to make better use of these tools as a means for maximizing protection of cultural resources and natural habitats.

2. If tribes see it as being in their best interests, undertake collaborative mapping of off-reservation TCPs, ranking those with multicultural significance as well as those facing impending threats. Attempt to use all legal designations possible to ensure maximum protection of these places under the Antiquities Act and the American Indian Religious Freedom Act.

3. In retreat settings, host tribal chairpersons, natural and cultural resource department heads, and cultural preservation specialists to brief them on their options for protecting TCPs. Tribal leaders may be eager to learn about innovative land designation and purchase strategies by conservation organizations, and can describe their own conservation priorities and the means of protection most acceptable to their communities.

4. To the extent that it fits within the interests of two or more tribes, foster collaboration with the National Register of Historic Landmarks to designate additional areas, and to gain cultural access agreements to off-reservation places, thus helping to maintain the living connections between indigenous people and their sacred places.

5. Advance more nominations of biologically diverse, culturally significant habitats as agricultural, subsistence, or cultural landscape sites for protection as NHLs or NNLs.

6. Involve the Trust for Public Lands' Tribal Lands Program in efforts to purchase TCPs under threat on the Colorado Plateau for traditional tribal uses, including gathering of sacred plants.

7. Work to educate all of the region's residents about the differences between how each culture relates to the land, and find ways to bridge these gulfs, helping to bring about cross-cultural understanding and collaboration.

Only through ongoing dialogue can we hope to develop the mutual respect and trust required for a true partnership. If we are successful, the strength of what we have begun may help foster other partnerships between federal agencies and Indian tribes.

—Jan R. Balsom (2001)

Living As if Biocultural Diversity Matters: Conservation Opportunities and Recommended Actions

Gary Paul Nabhan and David Seibert

An underlying premise of this report is that neither species nor languages can be safeguarded for long if conservation efforts focus on small, isolated fragments; both wildlife and vibrant human cultures need large tracts of "safe" habitat to survive and thrive. Large landscape-based conservation plans, to date, have focused scientific analysis and policy on (1) patterns of species richness on continents or within ecoregions (Mittermeier et al. 1998); (2) current distributions of protected areas and roadless zones that could be better managed for biodiversity, linked through corridors across private or commons lands (Soulé and Terborgh 1999); or (3) distributions of rare, threatened, and endangered species and unique habitats (Stein et al. 2000). In all three cases, the information, values, and priorities used to shape these plans come primarily, if not exclusively, from the Western scientific community of conservation biologists. In contrast, we have attempted, through literature reviews, workshops, and surveys, to catch a wider variety of cultural perspectives on our region, and to broaden the conservation plan focus from biodiversity to biocultural diversity.

If we listen and absorb the many voices and perspectives offered in the previous chapters, we can be dazzled by the many unique natural and cultural features that are juxtaposed on the Colorado Plateau—schists and shales, condors and cryptogamic crusts, hanging gardens and Zuni waffle gardens, monolingual Navajo farmers and multilingual Basque sheepherders. Taking it all in, it is all too easy to become frustrated if not immobilized by the range and the severity of threats, and by the rapidity of environmental and cultural change already apparent in this ecoregion. Given all that has been presented here—in all its complexity and with so much uncertainty apparent—we must take care not to lose sight of the questions that drive our work: What can our communities do to safeguard this uniqueness? What actions can we take to stave off the multiple threats to biocultural diversity on the Colorado Plateau? And what is the best course of action we can collectively take to ensure that the many perspectives offered here are adequately taken into account?

We must always ask such questions, or we will face the possibility that the work we have collected here, as well as what we may do collaboratively in the future, will amount to little more than academic exercises. But we must also ask questions regarding the decision-making processes that affect natural and cultural resources, especially with regard to their implications for environmental justice. In this realm, we see value in several of the *en libra* principles being actively debated in the Intermountain West (www.westgov.org):

- National standards, neighborhood solutions—assign responsibilities at the right level.
- Collaboration, not polarization—use collaborative processes to break down barriers and find solutions.

- Reward results, not programs—move to a performance-based system.
- Science for facts, process for priorities—separate subjective choices from objective data gathering.
- Change a heart, change a nation—environmental understanding is crucial.
- Recognition of costs and benefits—make sure environmental decisions are fully informed.
- Solutions transcend political boundaries—use appropriate geographic boundaries for environmental problems.

The ecoregional profile offered here and the dialogues from which it is derived are somewhat valuable in and of themselves, but there is even greater potential to put ideas into practice on the Colorado Plateau through collaborative restoration and conservation initiatives. Much remains to be accomplished, and our guess is that if it is to be accomplished, it will be through collaborative, community-based efforts informed by the synergies among Western and indigenous science (O'Neill 2000; Nabhan et al. 2002).

> After all, who wants the land stripped of its biological diversity? At the same time, who is not a land user? In reality, the vast majority of us are both "environmentalists" and "resource users." The problem is more about how to get things done. This I can assure you: if enough caring people make wildlife conservation a genuine part of everyday life and constructively engage friends, neighbors, and business associates, today's conflicts over endangered species will dissipate as quickly as an ephemeral thunderstorm on a summer afternoon high in the San Juan Mountains.
>
> —Tony Povilitis (2000)

There is broad consensus among those with whom we have interacted to compile this ecoregional profile: we wish to restore the integrity and health of the Colorado Plateau, and minimize the future effects of disruptive economic and political forces that lead to fragmentation of cultural and natural communities. Conditions on the Colorado Plateau today, for better and worse, are the direct results of multiple forces—many of them emanating from far beyond this region—which have dovetailed with one another to create change, some of it irreversible, some not. In essence, conservation biologists, ecological restorationists, and language revitalization activists assume that some kinds of change and the rapid rates at which they occur are detrimental to life; they believe that ecological and cultural integrity are valuable assets for all of us who live in this region. We must continue to value unique habitats, species, cultures, and languages, and recognize how they enrich us. In other words, we are concerned about the increasing homogenization of habitats and languages, and we wish to slow or reverse those trends while educating the public about the issues and how they might be addressed at the local level.

There also remains considerable polarization, especially with regard to the tolerable levels of urban growth, environmental contamination from mining, rangeland degradation from poor grazing practices, and the need for more protected areas where recreational and extractive uses are prohibited. We are also humbled by the gaps in information that remain with regard to the present levels of extant biodiversity, the extent of competition from invasive species, the degree of persistence of indigenous agriculture, and the relative intactness of cultural traditions. Nevertheless, we do not always need to wait until we have comprehensive, "perfect" knowledge to initiate an effective and cost-efficient conservation action.

There is great potential here on the Colorado Plateau—as in other communities and regions—to create and maintain a vision of lasting value, a vision that creates optimism in the face of obstacles. This optimism can be further fostered by community-based restoration efforts—innovations that advance environmental and social sustainability by providing novel solutions to cultural, economic, and ecological problems (Propst and Culp 2000). We recognize that the health of the land is of concern to both ecological restorationists and cultural treasure-keepers, whose very vocabularies are rich in terms describing the land and how it heals itself.

Cultural health is as integral to this vision as ecological health, and we find many connections between the two "encoded" in native languages (Nabhan et al. 2002).

We are committed to saving the remaining pieces of habitats and languages. As restorationists, we are interested in the historic reference conditions of natural and agricultural habitats and native language communities on the Colorado Plateau, but pragmatically recognize that we may not be able to return them to historic conditions in all cases; instead we will try to guide them on a trajectory that takes into account recent and anticipated climate change. We recognize that we cannot keep plant and animal populations or languages in some isolated, pure and pristine form. Instead, we may need to use novel interventions to revitalize them.

Collaborations across administrative boundaries, cultures, and disciplines seem to us to be critical to implementing this vision, as are community-based solutions. We hope to foster a deeper sense of place, with the understanding that just as endemic species belong to particular biotic communities, so do languages belong to specific cultural and environmental contexts. Active, informed, collaborative work can mitigate the effects of past mistakes to habitats and to cultures. The success of this kind of work depends in part on listening to and involving a broader spectrum of community members to assess the whole gamut of goals, practices, and mechanisms that they have developed in place, within a given cultural and ecological region. Ecosystems, cultures, and languages all change in discernible ways, in patterns that interconnect them. We have seen how forests and other landscapes have some capacity to heal themselves naturally; parallel processes may occur with cultures and with language change.

All of our initiatives should take care to save the remaining pieces and build on the natural or traditional processes that remain and function in place. Before attempting directed restoration of habitats or languages, for example, we should study past attempts that have failed, and learn from them. We should take into account the local context and pace of each situation and how it might be vulnerable or resilient to various pressures.

Given this broad biocultural perspective for future action to safeguard and restore the uniqueness of the Colorado Plateau, we endorse the recommendations elaborated by representatives from 12 of the indigenous and immigrant cultures of the region who participated with us in two workshops held at the Arboretum of Flagstaff, one in June of 2001 and one in March of 2002. In other words, we believe that the integration of tribal and nontribal efforts could result in the most dramatic advances possible for biocultural conservation:

1. We should form a multicultural consortium of scientists and other professionals to ensure that unique natural, agricultural, and cultural landscapes and waterscapes remain protected from new and impending threats, rather than having to invest in costly restoration after damage has been done.

2. We should foster and ensure funding for reciprocal exchanges among professionals from various cultures who inventory, monitor, and restore these landscapes and waterscapes, to promote cross-training in the use of comparable methods for undertaking inventories, for monitoring, and for advancing the best practices for restoration and recovery.

3. We should adopt as policy the National Council for Science and Environment recommendation that indigenous peoples' values be integrated into science to sustain the environment by recognizing that indigenous peoples have pioneered alternative ways of protecting habitats and their wildlife; indigenous science and values should carry the same weight and importance as Western academic science in decision making to promote the best available stewardship practices.

4. We should adopt as our mission the fostering of partnerships between scientists and community elders who maintain the traditional ecological knowledge of their cultures about plants, animals, springs, wetlands, wildlife, and agriculture.

5. We should adopt as our public service to communities our role in bridging the current information between technically oriented professionals and the communities who depend upon threatened aquifers, springs, rivers, and lakes to sustain their everyday lives. We should make it a practice to translate technical information into terms comprehensible to community members, who should participate in public hearings.

6. We should ensure that natural and cultural habitats are used as sites for environmental education, and that our programs offer internship opportunities for young tribal members to be exposed to careers in conservation and agricultural sciences.

7. We should use a diversity of means and media to inform communities about potential actions and threats to places and resources dear to them, and to get their perspectives on the management of these habitats and species, making sure that governments heed their mandate to ensure consultation with communities, even when it requires that it be done in native languages.

8. We should develop a watch-list of culturally sensitive plant and animal species associated with the Colorado Plateau that one or more tribes require for the persistence of their traditions, protected under the American Indian Religious Freedom Act. We should then collectively affirm that we have zero tolerance for their loss (due, for example, to groundwater pumping or land degradation) from any traditional cultural place, on reservations or on public lands.

9. We should attempt to pilot the use of alternative federal, tribal, and state protected area designations to legally protect traditional cultural places, and to maintain or revive knowledge, values, stories, and songs about them.

10. We should develop among tribal programs, universities, agencies, and nonprofit organizations a database of endemic plants and wildlife of the Colorado Plateau, listing sources of genetic resources available for restoration and recovery efforts.

11. We should establish an intertribal exchange of cultivated plant materials needed for agricultural revitalization, and protect these from patenting or commercial exploitation by outside interests.

12. We should establish an intertribal and interagency "SWAT" team which, with advance invitation, can assist tribal personnel with special techniques for removing the invasive species such as Russian olive and tamarisks that reduce the flow of springs; non-herbicidal methods should be promoted.

13. We should encourage all tribal natural resource, forestry, wetlands, and wildlife departments to consult frequently with cultural resource advisory teams before taking any actions that would affect sacred springs, sacred mountains, and other cultural landscapes.

14. We should encourage more frequent dialogue among multicultural nonprofit groups involved in environmental justice issues, and tribal programs, so that information is exchanged and tribal policies are understood.

15. We should encourage tribal interns and students to tape and write down in their own native languages the oral histories of places with particular cultural and historic significance to their communities.

16. We should ultimately work together toward developing an ecoregional conservation plan of protected areas, corridors, or buffers that takes into account unique natural and cultural resources; we should support one another in obtaining adequate funding from foundations and agencies to protect and restore these features.

17. We should support all efforts to ensure that the region's indigenous languages remain spoken in living communities by people of all ages, and that their vocabularies of the natural world are celebrated.

Appendix 1. Checklist of Native Mammals of the Colorado Plateau

Checklist of native mammals of the Colorado Plateau provided by Dr. David Armstrong, annotated for their known occurrence in national parks and monuments versus Indian reservations; 16 subspecies and two full species are endemic to the region.

	Res[1]	NPS[2]	Notes
INSECTIVORA SORICIDAE: SHREWS			
Masked Shrew, *Sorex cinereus*	—	—	Peripheral mountains
Merriam's Shrew, *Sorex merriami*	—	—	—
Montane Shrew, *Sorex monticolus*	X	X	Peripheral mountains
Dwarf Shrew, *Sorex nanus*	—	X	
Common Water Shrew, *Sorex palustris*	X	X	Peripheral mountains
Desert Shrew, *Notiosorex crawfordi*	X	—	
CHIROPTERA VESPERTILIONIDAE: COMMON BATS			
California Myotis, *Myotis californicus*	X	X	—
Western Small-footed Myotis, *Myotis ciliolabrum*	—	—	—
Longeared Myotis, *Myotis evotis*	X	X	—
Little Brown Bat, *Myotis lucifugus*	?	X	—
Fringed Myotis, *Myotis thysanodes*	X	X	—
Longlegged Myotis, *Myotis volans*	X	X	—
Yuma Myotis, *Myotis yumanensis*	X	X	—
Western Red Bat, *Lasiurus blossevillii*	—	—	Peripheral
Hoary Bat, *Lasiurus cinereus*	X	X	—
Silverhaired Bat, *Lasionycteris noctivagans*	—	—	—
Western Pipistrelle, *Pipistrellus hesperus*	X	X	—
Big Brown Bat, *Eptesicus fuscus*	X	X	—
Spotted Bat, *Euderma maculatum*	—	X	—
Townsend's Bigeared Bat, *Plecotus townsendii*	X	X	—
Allen's Bigeared Bat, *Idionycteris phyllotis*	X	—	—
Pallid Bat, *Antrozous pallidus*	X	X	—
MOLOSSIDAE: FREETAILED BATS			
Brazilian freetailed bat, *Tadarida brasiliensis*	X	X	—
Big Freetailed Bat, *Nyctinomops macrotis*	X	X	—
LAGOMORPHA OCHOTONIDAE: PIKAS			
Pika, *Ochotona princeps*	X	X	Peripheral
LEPORIDAE: RABBITS & HARES			
Desert Cottontail, *Sylvilagus audubonii*	X	X	—
Eastern Cottontail, *Sylvilagus floridanus*	X	X	Peripheral
Nuttall's Cottontail, *Sylvilagus nuttallii*	X	?	—
Blacktailed Jackrabbit, *Lepus californicus*	X	X	—
Whitetailed Jackrabbit, *Lepus townsendii*	X	X	—
RODENTIA SCIURIDAE: SQUIRRELS			
Cliff Chipmunk, *Tamias dorsalis*	X	X	—
Least Chipmunk, *Tamias minimus*	X	X	Peripheral mountains
Hopi Chipmunk, *Tamias rufus*	X	—	Eendemic to Hopi country
Uintah Chipmunk, *Tamias umbrinus*	—	X	Peripheral Henry Mtns only
Yellowbellied Marmot, *Marmota flaviventris*	—	—	Peripheral mountains
Whitetailed Antelope Squirrel, *Ammospermophilus leucurus*	X	X	—
Goldenmantled Ground Squirrel, *Spermophilus lateralis*	X	X	Peripheral mountains
Spotted Ground Squirrel, *Spermophilus spilosoma*	X	X	Endemic: *S. s. cryptospilotus* is endemic to Hopi and Navajo country
Rock Squirrel, *Spermophilus variegatus*	X	X	—
Gunnison's Prairiedog, *Cynomys gunnisoni*	X	X	Endemic: *C. g. zuniensis* is endemic to area around Zuni
Whitetailed Prairiedog, *Cynomys leucurus*	—	—	Peripheral
Utah Prairiedog, *Cynomys parvidens*	—	—	—

Appendix 1 (continued)

	Res[1]	NPS[2]	Notes
Abert's Squirrel, *Sciurus aberti*	X	X	Three endemic subspecies: *S. a. aberti* is broadly distributed; *S. a. kaibabensis* is endemic to N. Rim of Grand Canyon; and *S. a. chuskensis* is endemic to Navajo lands (Chuska & Tunitcha Mtns)
Pine Squirrel, or Chickaree, *Tamiasciurus hudsonicus*	X	X	Peripheral mountains
GEOMYIDAE: POCKET GOPHERS			
Botta's Pocket Gopher, *Thomomys bottae*	X	X	Endemic: *T. b. aureus* is endemic to Hopi and Navajo country
Northern Pocket Gopher, *Thomomys talpoides*	X	X	Peripheral mountains
HETEROMYIDAE: POCKET MICE AND ALLIES			
Arizona Pocket Mouse, *Perognathus amplus*	X	X	—
Plains Pocket Mouse, *Perognathus flavescens*	—	X	Endemic: *P. f. apache* is endemic to greater Western Apache country
Silky Pocket Mouse, *Perognathus flavus*	X	X	Endemic: *P. f. hopiensis* is endemic to Hopi country
Longtailed Pocket Mouse, *Perognathus formosus*	X	X	—
Little Pocket Mouse, *Perognathus longimembris*	X	X	—
Great Basin Pocket Mouse, *Perognathus parvus*	X	X	Endemic: *P. p. trumbullensis* is endemic to greater Kaibab, S. Pauite country
Rock Pocket Mouse, *Chaetodipus intermedius*	X	X	—
Chiseltoothed Kangaroo Rat, *Dipodomys microps*	X	X	—
Ord's Kangaroo Rat, *Dipodomys ordii*	X	X	Endemic: *D. o. longipes* is on reservation and park lands
Bannertailed Kangaroo Rat, *Dipodomys spectabilis*	X	—	Endemic: *D. s. clarencei* is found largely on reservation lands
CASTORIDAE BEAVERS			
American Beaver, *Castor canadensis*	X	X	Riparian
MURIDAE: RATS AND MICE			
Western Harvest Mouse, *Reithrodontomys megalotis*	X	X	—
Brush Mouse, *Peromyscus boylii*	X	X	—
Canyon Mouse, *Peromyscus crinitus*	X	X	Endemic: *P. c. auripectus*
Cactus Mouse, *Peromyscus eremicus*	X	X	—
Whitefooted Mouse, *Peromyscus leucopus*	—	X	Peripheral
Deer Mouse, *Peromyscus maniculatus*	X	X	—
Northern Rock Mouse, *Peromyscus nasutus*	X	X	—
Pinyon Mouse, *Peromyscus truei*	X	X	—
Northern Grasshopper Mouse, *Onychomys leucogaster*	X	X	Endemic *O. l. melanophrys* NW of Colorado R.; *O. l. pallescens* ranges from Colorado Plateau to San Luis Valley
Whitethroated Woodrat, *Neotoma albigula*	X	X	Endemic: *N. a. laplataensis*
Bushytailed Woodrat, *Neotoma cinerea*	X	—	*N. c. arizonae* endemic to Upper Colo. River drainage, including Hopi & Navajo lands
Arizona Woodrat, *Neotoma devia*	—	—	Endemic: *N. d. sanrafaeli*, *N. d. monstrabilis*
Desert Woodrat, *Neotoma lepida*	X	X	Peripheral Utah West Desert
Mexican Woodrat, *Neotoma mexicana*	X	X	Endemic: *N. m. inopinata*
Stephen's Woodrat, *Neotoma stephensi*	X	X	Endemic: *N. s. relicta* is endemic to Navajo lands
Longtailed Vole, *Microtus longicaudus*	X	X	Peripheral mountains
Mexican Vole, *Microtus mogollonensis*	X	—	Endemic, monotypic species
Montane Vole, *Microtus montanus*	X	—	Peripheral mountains
Meadow Vole, *Microtus pennsylvanicus*	—	—	Peripheral mountains
Common Muskrat, *Ondatra zibethicus*	X	X	Riparian

Appendix 1 (continued)

	Res[1]	NPS[2]	Notes
ZAPODIDAE: JUMPING MICE			
Meadow Jumping Mouse, *Zapus hudsonius*	X	X	Peripheral is on Jemez Pueblo & White Mtn Apache country
Western Jumping Mouse, *Zapus princeps*	—	—	Peripheral mountains
ERETHIZONTIDAE: PORCUPINES			
Common Porcupine, *Erethizon dorsatum*	X	X	—
CARNIVORA CANIDAE: DOGS AND ALLIES			
Coyote, *Canis latrans*	X	X	—
Gray Wolf, *Canis lupus*	(X)	(X)	Formerly present on res. lands
Kit Fox, *Vulpes macrotis*	X	X	—
Red Fox, *Vulpes vulpes*	X	X	—
Gray Fox, *Urocyon cinereoargenteus*	X	X	—
URSIDAE: BEARS			
Black Bear, *Ursus americanus*	X	X	—
Grizzly Bear, *Ursus arctos*	(X)	(X)	Formerly present on res. lands
Ringtail, *Bassariscus astutus*	X	X	—
Raccoon, *Procyon lotor*	X	X	Rriparian, H commensal
MUSTELIDAE: WEASELS AND ALLIES			
Longtailed Weasel, *Mustela frenata*	X	X	—
Blackfooted Ferret, *Mustela nigripes*	(X)	(X)	—
American Mink, *Mustela vison*	—	—	Riparian
American Badger, *Taxidea taxus*	X	X	—
River Otter, *Lontra canadensis*	(X)	(X)	Riparian
MEPHITIDAE: SKUNKS			
Western Spotted Skunk, *Spilogale gracilis*	X	X	—
Striped Skunk, *Mephitis mephitis*	X	X	—
FELIDAE: CATS			
Mountain Lion, *Felis concolor*	X	X	—
Bobcat, *Lynx rufus*	X	X	—
ARTIODACTYLA CERVIDAE: DEER			
Elk, or Wapiti, *Cervus elaphus*	X	X	Reintroduced, mostly in mnts
Mule Deer, *Odocoileus hemionus*	X	X	—
ANTILOCAPRIDAE: PRONGHORN			
Pronghorn, *Antilocapra americana*	X	X	Reintroduced
BOVIDAE: CATTLE AND ALLIES			
Mountain, or Bighorn Sheep, *Ovis canadensis*	?	X	—

[1]On Indian reservation lands.
[2]In U.S. National Parks.
Vernacular and scientific names mostly follow Jones et al. (1997), except that *Microtus mogollonensis* is treated as a species distinct from *M. mexicanus*, following various authors (see Frey and LaRue 1993).

APPENDIX 2. ENDEMIC PLANTS OF THE COLORADO PLATEAU

Endemic plants of the Colorado Plateau.

Scientific Name	State and Common Name	Indian (NN) Lands	Used
Agavaceae			
Agave species nova	Deer Creek agave	AZ	X
Agave utahensis ssp. *kaibabensis*	Kaibab agave	AZ UT	X
Yucca angustissima var. *angustissima*	Fine-leaf yucca	AZ UT NM, NN	X
Yucca angustissima var. *kanabensis*	Kanab fine-leaved yucca	AZ UT	X
Yucca angustissima var. *toftiae*	Toft's fine-leaved yucca	UT	X
Apiaceae			
Aletes macdougallii var. *macdougalli*	MacDougal's aletes	AZ NN	–
Aletes macdougallii var. *breviradiatus*	Mesa Verde aletes	CO NM UT	–
Aletes sessiliflorus	Sessile-flowered aletes	AZ CO NN NM	–
Angelica wheeleri	Utah angelica	UT	–
Cymopterus acaulis var. *higginsii*	Higgins spring-parsley	NNUT	X
Cympoterus duchensis	Uintah Basin spring-parsley	CO UT	X
Cymopterus minimus	Least spring-parsley	UT	X
Lomatium concinnum	Colorado desert-parsley	CO	X
Lomatium junceum	Rush desert-parsley	AZ UT	X
Lomatium latilobum	Canyonlands desert-parsley	COUT	X
Lomatium scrabrum var. *tripinnatum*	Rough desert-parsley	AZ UT	X
Lomatium (Oreoxis) troteri	Trotter's spring-parsley		X
Lomatium graveolens	Zion desert-parsley	UT	X
Apocynaceae			
Cycladenia jonesii	Jones cycladenia	UT AZ	–
Asclepiadaceae			
Asclepias cutleri	Cutler's milkweed	UT AZ NN	X
Asclepias welshii	Welsh's milkweed	AZ NN UT	X
Asteraceae			
Agoseris glauca var. *cronquistii*	Pale false dandelion	?	–
Ambrosia sandersoni		?	X
Artemesia campestris var. *petiolata*	Moon Lake sagewort	UT	X
Artemesia norvegica var., *picetorium*		?	X
*Artemesia tridentata*var. *parciflora*	Small-flowered sage	?	X
Aster welshii	Welsh's aster	UT	–
Chrysothamnus nauseosus ssp arenarius	Rabbitbrush	AZ UT NM	X
Chrysothamnus nauseosus ssp. *iridis*	Rabbitbrush	UT	X
Chrysothamnus nauseosus ssp. *latisquameus*	Rabbitbrush	AZ NM	X
Chrysothamnus nauseosus ssp. *nitidus*	Rabbitbrush	AZ UT NM	X
Chrysothamnus nauseosus ssp. *psilocarpus*	Huntington Canyon rabbitbrush	UT	X
Chrysothamnus nauseosus ssp. *salicifolius*	Rabbitbrush	UT	X
Chrysothamnus nauseosus ssp. *uintahensis*	Rabbitbrush	UT	X
Cirsium barnebyi	Barneby's thistle	UT WY CO	X
Cirsium eatonii var. *harrisonii*	Tushar Montains thistle	UT	X
Cirsium murdockii	Murdock's thistle	UT	X
Cirsium owenbyi		?	X
Cirsium scariousum		?	X
Cirsium virginense	Virgin thistle	AZ UT NV	X
Ericameria aremeroides var. *gramineus*		?	–
Ericameria lignumviridis	Greenwood goldenbush	UT	–
Ericameria zionus	Cedar Breaks goldenbush	UT	–
Erigeron acomanus	Acoma fleabane	NM NN	–
Erigeron awapensis	Awapa daisy	UT	X
Erigeron bistiensis	Bisti fleabane	NM NN	X
Erigeron carringtoniae	Jane Carrington's daisy	UT	X
Erigeron kachinensis	Kachina daisy	CO UT	X
Erigeron maguirei var. *harrisoni*	Maguire's daisy	UT	X
Erigeron maguirei var. *maguirei*	Maguire's daisy	UT	X
Erigeron mancus	La Sal Mountains daisy	CO UT	X
Erigeron proselyticus	Cedar Canyon daisy	UT	X
Erigeron rhizomatus	Zuni fleabane	AZ NM NN	X
Erigeron sivinskii	Sivinski's fleabane	AZ NM NN	X
Erigeron wawahensis	Daisy	?	X
Erigeron zothecinus	Alcove daisy	UT	X

Appendix 2 (continued)

Scientific Name	State and Common Name	Indian (NN) Lands	Used
Flaveria mcdougalli	Grand Canyon fleabane	AZ NN	–
Errazurzia rotundata	Round dunebroom	AZNN	X
Gaillardia spathulata	Western blanketflower	CO UT	X
Grindelia laciniata	Monticello gumweed	AZNM NM UT	X
Helianthus anomalus	Anomalous sunflower	AZ NN	X
Helianthus deserticola	Desert sunflower	AZ UT	X
Heliomeris soliceps	Barneby's goldeneye	UT	–
Heterotheca jonesii	Jones' golden-aster	UT	X
Hymenoxys acaulis var. *nana*	cushion golden-flower	UT	X
Hymenoxys subintegra	Bitterweed	AZ	X
Isocoma humilis	Leverich's goldenbush	?	X
Lygodesmia arizonica	Arizona skeleton-plant	AZ CO NM UT	X
Lygodesmia doloresensis	Dolores River skeleton-plant	CO	X
Lygodesmia entrada	Entrada skeletonplant	UT	X
Lygodesmia grandiflora	Entrada skeleton-weed	?	–
Machaeranthera pinnatifida var. *paradoxa*		AZ CO UT	–
Perityle specuicola	Alcove rock-daisy	?	–
Platyschkuhria integrifolia var. *oblongifolia*	No common name	?	–
Senecio actinella	Flagstaff groundsel	AZ NM	X
Senecio castoreus	Groundsel	UT	X
Senecio cynthioides	White Mountain grounsel	AZ NM NN	X
Senecio franscicanus	San Francisco Peaks grounsel	AZ	X
Senecio fremontii var. *inexpectatus*	Fremont ragwort	UT	X
Stephanomeria tenuifolia var. *utahensis*	Narrow-leaved skeleton-plant	UT	–
Thelesperma subnudum var. *alpinum*	Border golden-thread	UT	X
Townsendia aprica	Last chance townsendia	AZ, UT	X
Townsendia smithii	Black Rock daisy	AZ	X
Xylorhiza cronquistii	Cronquist's wood-aster	UT	–
Xylorhiza glabriuscula var. *linearfolia*	Wood-aster	UT	–
Boraginaceae			
Cryptantha cinerea var. *pustulosa*	Pustulose cryptantha	AZ CO NMUT	–
Cryptantha creutzfeldtii	Creutzfeldt-flower	UT	–
Cryptantha elata	Cliffdweller's candlestick cryptantha	UT CO	–
Cryptantha flava	Plateau yellow cats-eye	AZ CONM UT WY	–
Cryptantha johnstonii	Large-flowered cryptantha	UT	–
Cryptantha mensana	Carbon cryptantha	CO UT	–
Cryptantha ochroleuca	Red Canyon cryptantha	UT	–
Cryptantha osterhouti	Osterhout cryptantha	AZ CO UT	–
Cryptantha paradoxa	Handsome cats-eye	NM NN UT	–
Cryptantha semiglabra	Pipe Springs cryptantha	UT	–
Hackelia gracilenta	Mesa Verde stickseed	CO NN	–
Hackelia ibapensis	Ibapah stickseed	UT	–
Oreocarya aperta	Grand Junction cats-eye	CO	–
Oreocarya rollinsii	Rollins cryptanth	CO UT WY	–
Physaria stylosa	Duchesne River twinpod	UT	–
Thelypodiopsis ambigua	Long Valley tumble-mustard	AZ UT	–
Thelypodiopsis aurea	Durango tumble-mustard	CO NM UT	–
Brassicaceae			
Arabis gracilipes	Flagstaff rockcress	AZ NV UT	–
Draba ramulosa	Creeping draba	UT	–
Draba sobolifera	Tushar Mountains draba	UT	–
Glaucophyllum suffretescens	Shrubby reed-mustard	UT	–
Lepidium alyssoides var. *junceum*	Lee's Ferry pepper-wort	UT	X
Lepidium intergrifolium var. *heterophyllum*	Cedar Canyon pepper-wort	UT	X
Lepidium montanum var. *claronense*	Casto Canyon pepper-wort	UT	X
Lepidium montanum var. *neeseae*	Neese's pepper-wort	UT	X
Lesquerella hemiphysaria var. *lucens*	Range Creek bladderpod	UT	–
Lesquerella kaibabensis	Kaibab bladderpod	AZ	–
Lesquerella navajoensis	Navajo bladderpod	NM NN	–
Lesquerella pruinosa	Frosty bladderpod	CO	–
Lesquerella tumulosa	Kodachrome Basin bladderpod	UT	–

Appendix 2 (continued)

Scientific Name	State and Common Name	Indian (NN) Lands	Used
Physaria acutifolia	Book Cliffs twinpod	?	–
Physaria grahamii	Graham's twinpod	?	–
Physaria newberryi var. *racemosa*	Bloomington twinpod	?	–
Physaria repanda	Indian Canyon twinpod	?	–
Physaria stylosa	Long-styled twinpod	?	–
Schoenocrambe argillacea	Clay reed-mustard	UT	–
Schoenocrambe barnebyi	Barneby's reed-mustard	UT	–
Cactaceae			
Coryphantha vivipera var. *neomexicana*	Spiny stars	NM NN	–
Echinocereus triglochidiatus var. *triglochidiatus*	Hedgehog cactus	AZ CO NM UT	–
Opuntia basilaris var. *heilii*	Heil's beavertail-cactus	UT	X
Opuntia basilaris var. *longireolata*	Beavertail prickly pear	AZ	X
Opuntia nicholii	Navajo Bridge beavertail prickly pear	AZ NN UT	X
Pediocactus bradyi	Brady's cactus	AZ NN	–
Pediocactus despainii	San Rafael pincushion-cactus	UT	–
Pediocactus knowltoni	Knowlton's cactus	NM CO	–
Pediocactus paradineii	Paradine plains cactus	AZ NN	–
Pediocactus peeblesianus var. *fickeisenae*	Fickeisen's pincushion cactus	AZ NN	–
Pediocactus peeblesianus var. *peeblesianus*	Peeble's cactus	AZ NN	–
Pediocactus sileri	Siler's pincushion- cactus	AZ UT	–
Pediocactus simpsoni var. *minor*	Simpson's cactus	CO UT WY	–
Pediocactus winkleri	Winkler's pincushion cactus	UT	–
Sclerocactus cloverii ssp. *brackiae*	Brack's fishhook cactus	NM NN	–
Sclerocactus contortus	Canyonlands fish-hook cactus	UT	–
Sclerocactus glaucus	Uintah Basin cactus	COUT	–
Sclerocactus mesae-verde	Mesa Verde cactus	CO NMNN	–
Sclerocactus terrae-canyonae	Fishhook cactus	UT	–
Sclerocactus whipplei var. *heilii*	Hard Wall Cactus	AZ NM UT	–
Sclerocactus whipplei var. *reevesei*	Hard Wall Cactus	AZ NM UT	–
Sclerocactus wrightii	Wright fishhook cactus	UT	–
Campanulaceae			
Silene rectinacea		?	–
Cleomella palmerana var. *goodrichii*		UT CO	–
Capparaceae			
Cleomella palmeriana var. *goodrichii*		UT	–
Caryophyllaceae			
Arenaria abberrans	Sandwort	AZ	–
Silene rectiamea	Grand Canyon catch-fly	AZ	–
Chenopodiaceae			
Atriplex canescens var. *gigantea*	Dune saltbush	UT	X
Ceratoides lanata var. *ruinina*	Ruin Park winter-fat	UT	X
Proatriplex pleiantha			–
Cyperaceae			
Carex haysii	Zion sedge	UT	X
Carex scoparia	Sedge	AZ	X
Carex specuicola	Navajo sedge	AZ NN UT	X
Euphorbiaceae			
Euphorbia aaron-rosii			X
Euphorbia nephradenia	Utah spurge	UT	X
Fabaceae			
Astragalus accumbens	Zuni	NM NN	X
Astragalus ampullarioides	Shivits milk-vetch	?	X
Astragalus atwoodii	Milk-vetch	?	X
Astragulus barnebyi	Barneby's milk-vetch	AZ NN	X
Astragalus beathii	Beath's milk-vetch	AZ NN	X
Astragulus coltoni var. *moabensis*	Moab milk-vetch	AZUT	X
Astragalus cutleri	Cutler's milk-vetch	?	X
Astragalus cremnophylax var. *cremnophylax*	Milk-vetch	AZ	–
Astragalus cremnophylax var. *hevronii*	Hevron's century milk-vetch	AZ NN	X
Astragalus cremnophylax var. *myriorrhapis*	Century milk-vetch	AZ	X
Astragalus cronquistii	Cronquist's milk-vetch	CONNUT	X

Appendix 2 (continued)

Scientific Name	State and Common Name	Indian (NN) Lands	Used
Astragalus deterior	Milk	?	X
Astragalus equisolensis	Horseshoe Bend Milk-vetch		–
Astragalus eremiticus var. *ampullarioides*	Horseshoe Bend milk-vetch		X
Astragalus hamiltonii	Hamilton's milk- vetch	UT NM CO	X
Astragalus holmgreniorum	Pair-o-docs milk-vetch	AZ UT	X
Astragalus humillinus	Milk	?	X
Astragalus laccoliticus	Caineville milk-vetch	UT	X
Astragalus limnocharis var. *limnocharis*	Navajo Lake milk-vetch	UT	X
Astragalus limnocharis var. *tabulaeus*	Table Cliff Plateau milk-vetch	UT	X
Astragalus loanus	Glenwood milk-vetch	UT	X
Astragalus micromeris	Chaco milk-vetch	NM	X
Astragalus mollissimus	Mathew's wooly milk-vetch	NM	X
Astragalus moenkopensis	Moenkopi milk-vetch	AZ NN	X
Astragalus montii	Mont Lewis milk-vetch	UT	X
Astragalus naturitensis	Naturita milk-vetch	CONM NN UT	X
Astragalus newburyi var. *escalantinus*	Escalante milk-vetch	UT	X
Astragalus piscator	Fisher Towers milk-vetch	UT CO	X
Astragalus rafaelensis	Milk-vetch	UT	X
Astragalus rusbyi	Rusby's milk-vetch	AZ	X
Astragalus sabulosus var.*sabulosus*	Cisco milk-vetch	UT	X
Astragalus sabulosus var. *vehiculus*	Cisco milk-vetch	UT	X
Astragalus schmolliae	Milk-vetch	?	X
Astragalus serpens	Loa Pass milk-vetch	UT	X
Astragalus sesquiflorus	Milk-vetch	?	X
Astragalus sophoroides	Painted Desert milk-vetch	AZ NN	X
Astragalus striatiflorus	Sand milk-vetch	UT	X
Astragalus tortipes	Milk-vetch	NN	X
Astragalus welshii	Welsh's milk-vetch	UT	X
Astragalus xiphoides	Gladiator milk-vetch	AZ NN	X
Astragalus zionis var. *vigulus*	Browse milk-vetch	UT	X
Dalea flavescens var. *epica*		?	X
Lupinus crassus		?	X
Lupinus latifolius	Springdale lupine	UT	X
Pediomelum aromaticum	Paradox bread-root	?	–
Pediomelum aromaticum var. *barnebyi*	Barneby's bread-root	?	–
Pediomelum aromaticum var. *tuhyi*	Tuhy's bread-root	?	–
Pediomelum epipsilum	Kanab bread-root	?	–
Psoralea juncea	Scurf pea	AZ	–
Psorothamnus arborescens var. *pubescens*	House Rock Valley indigo-bush		X
Psorothamnus polydenius	Jones' indigo-bush		X
Psorothamnus arborescens var. *pubescens*	Marble Canyon dalea	AZ NN	X
Psorothamnus thompsoniae var. *whitingi*	Wupatki indigo-bush		X
Trifolium nuerophyllum	Mogollon clover	AZ NM	–
Hydrophyllaceae			
Nama retrorsum			–
Phacelia argillacea	Clay phacelia		–
Phacelia anelsonii	Aven Nelson's phacelia		–
Phacelia cephalotes		AZ	–
Phacelia cronquistiana	Cronquist's phacelia		–
Phacelia filiformis		AZ	–
Phacelia howe			–
Phacelia mam			–
Phacelia pulchella var. *sabulonum*	Kaiparowits phacelia		–
Phacelia pulchella var. *atwoodii*	Atwood's pretty		–
Phacelia reaf			–
Phacelia serrata			–
Phacelia utahensis	Utah phacelia		–
Phacelia welshii	Welsh phacelia	NN	–
Labiatae			
Hedeoma diffusum	Flagstaff pennyroyal	AZ NN	X
Salvia dorrii var. *mearnsii*	Mearns' sage	AZ	X

Appendix 2 (continued)

Scientific Name	State and Common Name	Indian (NN) Lands	Used
Astragalus cutleri	Cutler's milk-vetch	NNUT	X
Astragalus debaquaeus	Milk-vetch	?	X
Astragalus desereticus	Deseret milk-vetch	?	X
Liliaceae			
Allium bigelovii			X
Allium geyeri var. *chatterleyi*	Chatterley's onion	UT CO	X
Iris pariensis			X
Loasaceae			
Mentzelia			X
Mentzelia			X
Mentzelia shultziorum	Shultzes' blazing star		X
Mentzelia multicaulis	Book Cliffs blazing-star		X
Mentzelia goodrichii	Goodrich's blazing-star		X
Malvaceae			
Spharalcea grossularieafolia			X
Spharalcea leptophylla ja	Scaly globe-mallow	AZ	X
Spharalcea psor			X
Najadaceae			
Najas caespitosa	Fish Lake water-nymph		–
Onagraceae			
Abronia			–
Abronia			–
Camissonia atwoodii	Atwood's evening primrose	NNUT	–
Camissonia exilis	Slender evening primrose	AZ UT	–
Camissonia gouldii	Gould's evening primrose	AZ UT	–
Camissonia specuicola var. *specuicola*		AZ NN	–
Oenothera caespitosa			X
Oenothera			X
Ophioglossaceae			
Botrychium paradoxum	Paradox moonwort	UT	–
Botrychium echo			–
Orchidaceae			
Platanthera zothecina	Alcove bog-orchid	AZ NNUT	–
Spiranthes diluvialis			–
Papaveraceae			
Arctomecon californica	Las Vegas bearclaw-poppy	AZ NN NV UT	–
Arctomecon humilus	Dwarf bearclaw-poppy	UT	–
Argemone arizonica	Roaring Springs prickle-poppy	AZ	–
Philadelphaceae			
Jamesia americana var. *zionis*	Zion cliff-bush	UT	–
Poaceae			
Puccinella parishii	Parish alkali grass	AZ NM	–
Polemoniaceae			
Gilia caespitosa	Wonderland Alice-flower	UT	X
Gilia formosa			X
Gilia latifolia var. *imperialis*	Cataract Canyon gilia	UT	X
Gilia polyantha			X
Gilia tenuis	Muusentuchit gilia	UT	X
Ipomopsis spicata ssp. *tridactyla*	Tushar Mountains gilia	UT	–
Phlox amabilis		AZ	–
Phlox austromontana var. *lutescens*	Rimrock phlox		–
Phlox carophylaa	Pagosa phlox	NM CO	–
Phlox cluteana	Navajo phlox	AZ NN	–
Polygalaceae			
Polygala rusbyi		AZ	–
Polygonaceae			
Eriogonum aretioides	Widtsoe buckwheat	UT	X
Eriogonum brevicaule var. *promiscuum*	Mount Bartles buckhwheat	UT	X
Eriogonum corymbosum var. *cronquistii*	Cronquist's buckwheat	UT	X
Eriogonum corymbosum var. *hylophilum*	Gate Canyon buckwheat	UT	X
Eriogonum corymbosum var. *matthewsiae*	Springdale buckwheat	UT	X

Appendix 2 (continued)

Scientific Name	State and Common Name	Indian (NN) Lands	Used
Eriogonum corymbosum var. *smithii*	Arthur Smith's buckwheat	UT	X
Eriogonum darrovii	Darrow's buckwheat	UT	X
Eriogonum heermannii var.*subspinosum*	Bulldog Knolls buckwheat	UT	X
Eriogonum jamesii var. *higginsii*	Higgins' buckwheat	UT	X
Eriogonum racemosum var. *nobile*	Bluff buckwheat	UT	X
Eriogonum scabrellum	Westwater buckwheat	UT	X
Portulacaceae			
Talinum			–
Talinum validulum	Tusayan flame flower	AZ	–
Ranunculaceae			
Anemone multifida var. *stylosa*	Fish Lake windflower	AZ UT	–
Aquilegia desertorum	Desert columbine	AZ NN UT	–
Aquilegia flavescens var. *rubicunda*	Link Trail columbine	UT	–
Aquilegia formosa var. *formosa*	Foster's columbine	AZUT	–
Aquilegia micrantha var. *mancosana*	Mancos columbine	CO	–
Aquilegia micrantha var. *micrantha*	Little-flowered columbine	AZ CO UT	–
Clematis hirsutissima var. *arizonica*	Arizona leather-flower	AZ NN	–
Ranunculus aestivalis	Autumn buttercup		–
Rhamnaceae			
Ceanothus greggii var. *franklinii*	Ben's buck-brush	AZ UT	–
Rosaceae			
Cercocarpus intricatus var. *villosus*	Mountain mahogany	AZ UT	X
Cordylanthus wrightii ssp. *kaibabensis*	Kaibab slender-leaf bird's-beak	AZ	–
Crataegus douglasii var. *duchesnensis*	Duchesne River hawthorn	UT	–
Potentilla angelliae	Boulder Mtn. cinquefoil		–
Rosa stellata var. *abyssa*	Grand Canyon rose	AZ NN	–
Salicaceae			
Salix arizonica	Arizona willow	AZ NN	X
Saxifragaceae			
Heuchera eastwoodiae	Alum root	AZ	–
Jamesia americana zio	Cliffbrush	AZ	–
Scrophulariaceae			
Besseya oblo			–
Castilleja aquariensis	Aquarius paint-brush	UT	–
Castilleja kaibabensis	Kaibab paint-brush	AZ	–
Castilleja mogollonica	White Mountains paint-brush	AZ NN	–
Castilleja parvula	Tushar Mountain paint-brush	UT	–
Castilleja patriotica	Patriot paint-brush		–
Castilleja revealii	Reveal's paint-brush	UT	–
Mimulus eastwoodiae	Eastwood's monkey flower		–
Penstemon abietinus	Fir-leaved beard-tongue	UT	X
Penstemon ammophilus	Canaan Mtn. sand-loving beard-tongue	2	X
Penstemon angustifolius var. *venosus*	Narrow-leaved beard-tongue	AZ NM UT	X
Penstemon angustifolius var. *vernalensis*	Vernal narrow-leaved beard-tongue	CO UT	X
Penstemon barbatus ssp. *trichader*	Beard-tongue	AZ CO NM UT	X
Penstemon bracteatus	Red Canyon beard-tongue	UT	X
Penstemon breviculus	Short-stem beard-tongue	AZ CO NM UT	X
Penstemon caespitosus var. *desertipicti*	Mat beard-tongue	AZ COUT	X
Penstemon clutei	A beard-tongue	AZ NN	X
Penstemon crandallii	Crandall's beard-tongue	CO NM UT	X
Penstemon distans	Mount Trumbull beard-tongue	AZ	X
Penstemon dolius var. *duchesnensis*	Duchesne beard-tongue	UT	X
Penstemon flowersii	Flowers' beard-tongue	NN UT	X
Penstemon grahamii	Graham's beard-tongue	CO UT	X
Penstemon goodrichii	Goodrich's beard-tongue	UT	X
Penstemon humilis ssp. *obtusifolius*	Lowly beard-tongue	UT	X
Penstemon leonardii var. *higginsii*	Higgins beard-tongue	UT	X
Penstemon lentus var. *lentus*		AZ CO NM UT	X
Penstemon liniarioides ssp. *liniarioides*		AZ CO NM	X
Penstemon liniarioides ssp. *maguirei*	MacGuire's penstemon	AZ NM	X
Penstemon marcusii	Castle Valley beard-tongue	UT	X

Appendix 2 (continued)

Scientific Name	State and Common Name	Indian (NN) Lands	Used
Penstemon navajoa	Navajo Mtn. penstemon	NN UT	X
Penstemon nudiflorus	Flagstaff penstemon	AZ	X
Penstemon oliganthus	Apache beard-tongue	AZ NN	X
Penstemon parvus	Aquarius penstemon	UT	X
Penstemon pinorum	Pinyon penstemon	UT	X
Penstemon pseudoputus	Kaibab beard-tongue	AZ UT	X
Penstemon scariosus	White River beard-tongue		X
Penstemon scariosus var. *cyanomontanus*	Blue Mtn. Beard-tongue	CO UT	X
Penstemon virgatus	Upright beard-tongue	AZ NM	X

LITERATURE CITED

Abbey, E. 1968. Desert solitaire. McGraw-Hill, New York.

Abbey, E. 1975. The monkey wrench gang. Lippincott, Philadelphia.

Abbey, E. 1977. The journey home: Some words in defense of the American West. Illustrations by Jim Stiles. Dutton, New York.

Albert, S., and T. Trimble. 2000. Beavers are partners in riparian restoration on the Zuni Indian Reservation. Ecological Restoration 18 (2): 87–96.

Allen, C. D., J. L. Bentancourt, and T. W. Swetnam. 1998. Landscape changes in the southwestern United States: Techniques, long-term data sets, and trends. In Perspectives on the land use history of North America: A context for understanding our changing environment, edited by T. D. Sisk, pp. 71–84. U.S. Geological Survey, Biological Resources Division, Biological Science Report USGS/BRD/BSR-1998-0003.

Armstrong, D. M. 1982. Mammals of the canyon country. Canyonlands Natural History Association, Moab, Utah.

Armstrong, D. M. 2002. Endemism in mammals of the Colorado Plateau. Draft analysis in preparation by the author. Department of Environmental, Population and Organism Biology, University of Colorado, Boulder.

Baars, D. L. 1983. Canyonlands country. Canyonlands Natural History Association, Moab, Utah.

Baars, D. 1989. The Colorado Plateau: A geological history. University of New Mexico Press, Albuquerque.

Babcock, R. S. 1990. Precambrian crystalline core. In Grand Canyon geology, edited by S. S. Beus and M. Morales, pp. 11–29. Oxford University Press, Oxford.

Balsom, J. R. 2001. Inclusion in NPS management at Grand Canyon: Tribal involvement and integration. In Crossing boundaries in parks and on public lands: Proceedings of the 11th Conference on Research and Resource Management, edited by D. Harmon. George Wright Society, Hancock, Michigan.

Basso, K. 1996. Wisdom sits in places: Landscape and language among the Western Apache. University of New Mexico Press, Albuquerque.

Belnap, J., and J. S. Gardner. 1993. Soils microstructure in soils of the Colorado Plateau: The role of *Microcoleus vaginatus*. Great Basin Naturalist 53: 40–47.

Berkes, F. 2000. Sacred ecology. Yale University Press, New Haven, Connecticut.

Beus, S. S., and M. Morales, Editors. 1990. Grand Canyon geology. Oxford University Press, Oxford.

Bingham, S., and J. Bingham. 1979. Navajo farming. Utah State University Press for Rough Rock School, Navajo Nation (Arizona).

Brown, D. E. 1982. Biotic communities of the American Southwest—United States and Mexico. Desert Plants 4(1–2): 1–342.

Brown, K. 1995. Four Corners: History, land, and people of the desert Southwest. Harper Collins, New York.

Brown, W. E. 2001. Overview of subsistence in Alaska. In Crossing boundaries in parks and on public lands: Proceedings of the 11th Conference on Research and Resource Management, edited by D. Harmon. George Wright Society, Hancock, Michigan.

Burnham, P. 2000. Indian country, God's country: Native Americans and the national parks. Island Press, Washington, D.C.

Carothers, S. W., and B. T. Brown. 1991. The Colorado River through Grand Canyon: Natural history and human change. University of Arizona Press, Tucson.

Carothers, S. W., and D. A. House. 2000. Decommissioning Glen Canyon Dam: The key to Colorado River ecosystem restoration and the recovery of endangered species? Arizona Law Review 42(2): 215–238.

Christman, C. J., D. P. Sponenberg, and D. E. Bixby. 1997. A rare breeds album of American livestock. American Minor Breeds Conservancy, Pittsboro, North Carolina.

Comstock, J. P., and J. R. Ehleringer. 1992. Plant adaptation in the Great Basin and Colorado Plateau. Great Basion Naturalist 52(3): 195–215.

Czech, B., P. R. Krausman, and P. K. Devers. 2000. Economic associations among causes of species endangerment in the United States. BioScience (July), pp. 593–601.

Davenport, D. C., Editor. 1978. Factors influencing usefulness of antitranspirants applied on phreatophytes to increase water supplies. California Water Resources Center. University of California Davis Contribution 176.

Emslie, S. 1981. Birds and prehistoric agriculture: The New Mexican pueblos. Human Ecology 4: 331–349.
Estes, J. 1999. How many indigenous languages are spoken in the United States? By how many speakers? National Clearinghouse for Bilingual Education. Retrieved from http://www.ncbe.gwu.edu/askncbe/faqs/20natlang.htm
Euler, R. C., G. J. Gumerman, T. N. V. Karlstrom, J. S. Dean, and R. H. Hevly. 1979. The Colorado Plateau: Cultural dynamics and paleoenvironments. Science 205: 1089–1101.
Felger, R. S., and G. P. Nabhan. 1976. Deceptive barrenness. Ceres 9(2): 34–39.
Ford, R. 1981. Gardening and farming before 1000: Patterns of pre-historic cultivation north of Mexico. Journal of Ethnobiology 1: 6–27.
Fradkin, P. 1984. A river no more. Island Press, Washington D.C.
Grissino-Mayer, H., T. W. Swetnam, and R. K. Adams. 1997. The rare, old-aged conifers of El Malpais: Their role in understanding climatic change in the American Southwest. In Natural history of El Malpais National Monument, edited by K. Mabery, pp. 155–161. Bulletin 156. New Mexico Bureau of Mines and Mineral Resources, Socorro.
Gulliford, A. 2000. Sacred objects and sacred places: Preserving tribal traditions. University Press of Colorado, Boulder.
Hamblin, W. K. 1990. Late Cenozoic lava dams in the western Grand Canyon. In Grand Canyon geology, edited by S. S. Beus and M. Morales, pp. 435–442. Oxford University Press, Oxford.
Hansen, E. M., F. K. Schwarz, and J. T. Riedel. 1977. Probable maximum precipitation estimates, Colorado River and Great Basin drainages. National Oceananic and Atmospheric Administration, Hydrometeorological Report No. 49.
Harjo, J. 1989. Secrets from the center of the world. University of Arizona Press, Tucson.
Harper, K. T., and J. R. Marble. 1988. The role for non-vascular plants in management of arid and semiarid rangeland. In Vegetation science applications for rangeland analysis and management, edited by P. T. Tuelier, pp. 135–169. Kluwer Academic, Dordrecht, Germany.
Harper, K. T., L. L. St. Clair, K. H. Thorine, and W. M. Hess. 1994. Natural history of the Colorado Plateau and Great Basin. University Press of Colorado, Boulder.
Heath, R. C. 1984. Ground-water regions of the United States. U.S. Geological Survey Water-Supply Paper 2242.
Hecox, W. E., and B. Ack. 1996. Charting the Colorado Plateau: An economic and demographic exploration. Grand Canyon Trust, Flagstaff, Arizona.
Hendricks, R., and J. P. Wilson, Editors. 1996. The Navajos in 1705: Roque Madrid's campaign journal. University of New Mexico Press, Albuquerque.
Hereford, R. 1984. Climate and ephemeral-stream processes: Twentieth century geomorphology and alluvial stratigraphy of the Little Colorado River, Arizona. Geological Society of America Bulletin 95: 654–668.
Heywood, V. H., Editor. 1995. Global biodiversity assessment. Cambridge, UK: Cambridge University Press/ UN Environment Programme.
Higgins, D. 2001. Language decline and stabilization: A narrative of the revitalization of the Hopi language. Master's thesis, Northern Arizona University, Flagstaff.
High Country News. 2000. Delicate arch. High Country News archives, Paonia Colorado. Retrieved online.
Hogan, L. 1992. Journeys and other stories. In Columbus and beyond: Views from Native Americans, edited by R. Jorgen. Southwest Parks and Monuments Association, Tucson, Arizona.
Iorns, W. V., C. H. Hembree, and G. L. Oakland. 1965. Water resources of the upper Colorado River basin—Technical report. U.S. Geological Survey Professional Paper 441.
Ingraham, N. L., K. Zukosky, and D. K. Kreamer. 2001. Application of stable isotopes to identify problems in large-scale water transfer in Grand Canyon National Park. Environmental Science and Technology 35: 1299–1302.
Inouye, D., B. Barr, K. Armitage, and B. Inouye. 2000. Climate change is affecting altitudinal migrants and hibernating species. Proceedings of the National Academy of Sciences 97(4): 1630–1633.
Johansen, J. R. 1993. Cryptogamic crusts of semiarid and arid lands of North America. Journal of Phycology 29: 140–147.
Kartesz, J., and A. Farstad. 1999. Multi-scale analysis of endemism of vascular plant species. In Terrestrial ecoregions of North America: A conservation assessment, by T. H. Ricketts, E. Dinerstein, D. M. Olson, and C. Loucks, pp. 51–55. Island Press, Washington, D.C.
Krauss, M. 1996. Status of Native American language endangerment. In Stabilizing indigenous languages, edited by Gina Cantoni, pp. 16–21. Center for Excellence in Education, Northern Arizona University, Flagstaff.
Linford, L. D. 2000. Navajo places: History, legend, landscape. University of Utah Press, Salt Lake City.
Maffi, L. 2001, Editor. On biocultural diversity: Linking language, knowledge and the environment. Smithsonian Institution Press, Washington, D.C.
Maffi, L., T. Skutnabb-Kangas, and J. Andrianarivo. 1999. Linguistic diversity. In Cultural and spiritual values of biodiversity, edited by Darrell Posey, pp. 21–57. Intermediate Technology Publications/ UN Environment Programme, London/Nairobi.
Maffi, L., T. Skutnabb-Kangas, and D. Harmon. Forthcoming. Sharing a world of difference: The earth's linguistic, cultural, and biological diversity. UNESCO, Paris.

Masse, D. L., and V. D. Adams. 1983. Polycyclic aromatic hydrocarbons: Are they a problem in processed oil shales. Utah Water Res. Lab Water Quality Series UWRL/Q-83/07, Logan.
McLeod, C., and M. Maynor. 2001. In the light of reverence. Bullfrog Films, Oley, Pennsylvania.
McKusick, C. R. 1986. Southwest Indian turkeys: Prehistory and comparative osteology. Southwest Bird Laboratory, Globe, Arizona.
Merriam, C. H., and Steineger, L. 1890. Results of a biological survey of the San Francisco Mountain region and the desert of the Little Colorado, Arizona. North American Fauna Report 3. U.S. Department of Agriculture, Division of Ornithology and Mammalia, Washington, D.C.
Milne, C. 1995. Sacred Places in North America. Stewart, Tabori and Chang, New York.
Minnis, P., and W. J. Elisens. 2000. Biodiversity in Native North America. University of Oklahoma Press, Norman.
Mittermeier, R. A., N. Myers, J. B. Thomsen, G. A. B. da Fonseca, and S. Olivieri. 1998. Biodiversity hotspots and major tropical wilderness areas: Approaches to setting conservation priorities. Conservation Biology 12(3): 516–520.
Momaday, N. S. 1976. The names: A memoir. Harper and Row, New York.
Momaday, N. S. 1995. Foreword to Navajo place names: An observer's guide, edited by Alan Wilson. Jeffrey Norton, Guilford-on-the-Green, Connecticut.
Morris, I. 1997. From the glittering world. University of Oklahoma Press, Norman.
Museum of Northern Arizona. 1989. Wilderness areas of the Colorado Plateau: A special issue. Plateau magazine.
Nabhan, G. P. 1985. Native crop diversity in arid America: Conservation of regional gene pools. Economic Botany 39(4): 387–399.
Nabhan, G. P. 1986. Native American crop diversity, genetic resource conservation, and the policy of neglect. Agriculture and Human Values 2(3): 14–21.
Nabhan, G. P. 1997. Cultures of habitat: On nature, culture, and story. Counterpoint, Washington, D.C.
Nabhan, G. P. 2000a. Interspecific relationships affecting endangered species recognized by O'odham and Comcaac cultures. Ecological Applications 10(5): 1288–1295.
Nabhan, G. P. 2000b. Biodiversity and linguistic diversity on the Colorado Plateau. Biodiversity Newsletter. The Consultative Group on Biological Diversity.
Nabhan, G. P., and C. Wilson 1995. Canyons of color. Tehabi Books/Harper-Collins West, San Francisco.
Nabhan, G. P., D. House, H. Suzan, W. Hodgson, L. Hernandez, and G. Malda. 1991. Conservation and use of rare plants by traditional cultures of the U.S./Mexico borderlands. In Biodiversity: Culture, conservation and ecodevelopment, edited by M. Oldfield and J. Alcorn, pp. 127–146. Westview Press, Boulder, Colorado.
Nabhan, G. P., P. Pynes, and T. Joe 2002. Where biological and cultural diversity converge: Safeguarding endemic species and languages on the Colorado Plateau. In Ethnobiology and biocultural diversity, by J. R. Stepp, F. S. Wyndham, and R. K. Zarger, pp. 61–71. International Society for Ethnobiology. University of Georgia Press, Athens.
National Park System Advisory Board. July 2001. Rethinking the national parks for the 21st century. U.S. Department of the Interior, Washington, D.C.
Natural Resources Defense Council. 2001. Groundwater mining of Black Mesa. NRDC, New York.
Nelson, C. R. 1994. Insects of the Great Basin and Colorado Plateau. In Natural history of the Colorado Plateau and Great Basin, edited by K. Harper et al., pp. 211–254. University Press of Colorado, Boulder.
O'Neill, C. A. 2000. Restoration affecting native resources: The place of native ecological science. Arizona Law Review 42(2): 343–380.
Parker, P., and T. King. 1990. Guidelines for evaluating and documenting traditional cultural properties. National Register Bulletin 38. U.S. Department of the Interior, National Park Service, Interagency Resource Division, Washington, D.C.
Pawluk, R. R., J. A. Sandor, and J. A. Tabor. 1992. The role of indigenous soil knowledge in agricultural development. Journal of Soil and Water Conservation 47: 298–302.
Pontius, D. 1997. Colorado River basin study final report. SWCA, Tucson.
Posey, D., Editor. 1999. Cultural and spiritual values of biodiversity. Intermediate Technology Publications/UN Environment Programme, London/Nairobi.
Povilitis, T. 2000. Slipping through our hands: Imperiled wildlife of the greater San Juans. Lifenet, Wilcox, Arizona.
Price, D., and T. Arnow. 1974. Summary appraisal of the nation's ground-water resources, Upper Colorado Region. U.S. Geological Survey Professional Paper 813-C.
Propst, C. L., and P. W. Culp. 2000. Searching for Ciboila: Community-based environmental restoration in the Colorado River watershed. Arizona Law Review 42(2): 259–286.
Rabinowitz, D., S. Cairns, and T. Dillon. 1986. Seven forms of rarity and their frequency in the floras of the British Isles. In Conservation biology: The science of scarcity and diversity, edited by M. E. Soulé, pp. 182–204. Sinauer Associates, Sunderland, Massachusetts.
Reisner, M. 1993. Cadillac desert. Penguin Books, New York.
Ricketts, T. H., E. Dinerstein, D. M. Olson and C. Loucks. 1999a. Who's where in North America? BioScience 49(5): 369–381.

Ricketts, T. H., E. Dinerstein, D. M. Olson, and C. Loucks 1999b. Terrestrial ecoregions of North America: A conservation assessment. Island Press, Washington, D.C.
Rykaczewski, D. 1981. Climate of the San Juan resource area. Bureau of Land Management, Salt Lake City.
Sellars, R. W. 1997. Preserving nature in the national parks: A history. Yale University Press, New Haven.
Shand, H. 1997. Human nature—Agricultural biodiversity and farm-based food security. Rural Advancement Foundation International. Retrieved online from www.etcgroup.org/documents/other_human.pdf
Sheridan, T. E., and G. P. Nabhan 1979. Traditional technology for floodplain management. In Proceedings of the Symposium on Flood Monitoring and Management of the Arizona Section of the American Water Resources Association, Tucson, October 26, pp. 73-81.
Sigler, J. W., and W. F. Sigler. 1994. Fishes of the Great Basin and Colorado Plateau. In Natural history of the Colorado Plateau and Great Basin, edited by K. Harper et al., pp. 163–210. University Press of Colorado, Boulder.
Silko, L. M. 1987. Landscape, history, and the Pueblo imagination. In On nature: Nature, landscape and natural history, edited by Daniel Halpern. North Point Press, San Francisco.
Sisk, T. D., A. E. Launer, K. R. Switky, and P. R. Ehrlich. 1994. Evaluating extinctions threats: The distribution of global biodiversity and the expansion of the human enterprise. BioScience 44: 592–604.
Smith, C., and E. Manning. 1997. The sacred and profane collide in the West. High Country News (May 26), p. 1.
Soulé, M. E., and J. Terborgh. 1999. Continental conservation: Scientific foundations of regional reserve networks. Island Press, Washington, D.C.
Spence, J., and N. Henderson. 1993. Tinaja and hanging garden vegetation in Capitol Reef National Park, southern Utah, USA. Journal of Arid Environments 24(1): 21–37.
Stein, B., L. Kutner, and J. Adams, Editors. 2000. Precious heritage: The status of biodiversity in the United States. Oxford University Press, Oxford.
Stevens, L. E., and T. J. Ayres. (in press). The biodiversity and distribution of alien vascular plants and animals in the Grand Canyon region. In Invasive species of the Sonoran Desert, edited by B. Tellman. University of Arizona Press, Tucson.
Stoffle, R. 1997. Waterscapes in American Indian ethnogeography. Paper presented at the annual meetings of the Society for Applied Anthropology. Seattle, Washington.
Stoker, T. N. 1990. Cultural and biological diversity at Canyon de Chelly National Monument, Arizona. Society and Natural Resources 3: 349–360.
Tapahonso, L. 1993. Saanii dahataal, the women are singing: Poems and stories. University of Arizona Press, Tucson.
Tapahonso, L. 1995. Ode to the land: The Diné view. New Mexico Magazine (August), pp. 62–69.
Thomas, H. E. 1952. Ground-water regions of the United States: Their storage facilities, the physical and economic foundation of natural resources. 83rd U.S. Congress, House Committee on Interior and Insular Affairs. v. 3: 3–78.
Trimble, S. 1991. Blessed by light: Visions of the Colorado Plateau. Gibbs Smith, Layton, Utah.
U.S. Department of Interior. 1987. Quality of water, Colorado River basin. U.S. Government Printing Office, Washington, D.C.
Utah Division of Natural Resources. 1998. Inventory of sensitive species and ecosystems in Utah. Retrieved online from www.utahcdc.usu.edu/ucdc/ViewReports/plantrpt.pdf
Van Dyke, J. 1920. The Grand Canyon of the Colorado: Recurrent studies in impressions and appearances. C. Scribner's Sons, New York.
Vasquez, M., and L. Jenkins Kuwanwisima. 1994. Reciprocity and sustainability: Terrace restoration on Third Mesa. Practicing Anthropology 16(2): 14–18.
Walka, A. 2002. Water from the rock. Plateau Journal 5(2): 7–21.
Wegner, D. L. 1998. Restoring Glen Canyon: Linking our future to the importance of our past. Glen Canyon Institute, Salt Lake City, Utah and Flagstaff, Arizona. Retrieved from http://www.glencanyon.org/facts/cea/wegnerover.htm
Wegner, D. L. 2000. Looking toward the future: The time has come to restore Glen Canyon. Arizona Law Review 42(2): 239–258.
Welsh, S. 1989. On the distribution of Utah's hanging gardens. Great Basin Naturalist 49(1): 1–30.
White, C. S., D. R. Dreesen, and S. R. Loftin. 1996. Cobble mulch: An Anasazi water-conservation gardening technique. Proceedings of the 42nd Annual New Mexico Water Conference, Albuquerque, New Mexico.
Whiting, A. F. 1942. The ethnobotany of the Hopi. Museum of Northern Arizona, Flagstaff.
Wilkinson, C. 1999. Fire on the plateau: Conflict and endurance in the American Southwest. Island Press, Washington, D.C.
Windell, J. T. 1986. Impacts to wetlands and implications for management of Rocky Mountain wetlands. In An ecological characterization of Rocky Mountain montane and subalpine wetlands, edited by J. T. Windell, B. E. Willard, D. J. Cooper, S. Q. Foster, C. F. Knud-Hansen, L. P. Rink, and G. N. Kiladis, pp. 215–236. U.S. Fish and Wildlife Service Report 86(11). National Ecology Center, Washington, D.C.

Windell, J. T., B. E. Willard, D. J. Cooper, S. Q. Foster, C. F. Knud-Hansen, L. P. Rink, and G. N. Kiladis. 1986. An ecological characterization of Rocky Mountain montane and subalpine wetlands. U.S. Fish and Wildlife Service Biol. Report 86(11).

Winter, J. C. 1974. Aboriginal agriculture in the Southwest and Great Basin. Ph.D. dissertation, University of Utah, Salt Lake City.

Woolsey, A. I. 1980. Agricultural diversity in the prehistoric Southwest. The Kiva 45: 317–336.

World Wildlife Fund International—Terralingua 2000. Indigenous and traditional peoples of the world and ecoregion conservation: An integrated approach to conserving the world's biological and cultural diversity. WWF-International, Gland, Switzerland.

Worster, D. 1985. Rivers of empire: Water, aridity and the growth of the American West. Pantheon, New York.

Working Group on the Endangered Species Act and Indian Water Rights. 2001. Case study on the upper Colorado Basin.

Yamamoto, A., M. S. Linn, T. Naranjo, S. Nicholas, I. Yang-Slaughter, and O. Zepeda. 2002. Awakening the languages: Field reports from 15 programs from Arizona, New Mexico and Oklahoma. In Indigenous languages across the community, edited by B. Burnaby and J. Reyhner, pp. 105–126. Center for Excellence in Education, Northern Arizona University, Flagstaff.

Web sites consulted

http://geochange.er.usgs.gov/sw/impacts/biology/crypto [USGS Global Change Research]

http://www.amrivers.org [American Rivers]

http://cpluhna.nau.edu [Canyons, Cultures and Environmental Change: an Introduction to the land use history of the Colorado Plateau]

http://www.itcaonline.com [Havasupai, Hualapai, Hopi, Kaibab Band of Paiute Indians, Yavapai Prescott Tribe]

http://www.carizona.com [Camp Verde Yavapai-Apache]

http://www.yavapai-apache-nation.com [Camp Verde Yavapai-Apache]

http://www.dced.state.ut.us/indian [Paiute Tribe of Utah]

http://www.southern-ute.nsn.us/government/language.html [Southern Ute Tribe]

http://www.ubtanet.com/~northernute [Northern Ute Tribe]

http://www.dced.state.ut.us/indian/Today/ute.html [Northern Ute Tribe]

http://epa.gov [Pueblo of Acoma]

http://www.zuni.k12.nm.us [Zuni Pueblo]

http://www.swr-hq.nm.bia.gov/ACRES-19.99.htm [Bureau of Indian Affairs: "Southwest Region Indian Acreages, as of December 31, 1999"]

http://www.swr.-hq.nm.bia.gov/99ABQlf.htm [Bureau of Indian Affairs: "Southwest Region Combined Totals"]

http://www.indigenous-language.org [Indigenous Languages Institute, Santa Fe, New Mexico: "Endangered Languages Database." Accessed December 5, 2001]

http://www.eri.nau.edu/FAQ.htm [Ecological Restoration Institute, Northern Arizona University. "Restorating the Southwestern Ponderosa Pine Forest." Accessed December 20, 2001]